The Panda's Thumb

More Reflections in Natural History

The Panda's Thumb

More Reflections in Natural History

Stephen Jay Gould

W · W · NORTON & COMPANY
NEW YORK   LONDON

First published as a Norton paperback 1982

Copyright © 1980 by Stephen Jay Gould
Published simultaneously in Canada by George J. McLeod
Limited, Toronto. Printed in the United States of America.
ALL RIGHTS RESERVED

Library of Congress Cataloging in Publication Data
Gould, Stephen Jay.
    The panda's thumb.
    Bibliography: p.
    Includes index.
    1. Evolution—History.   2. Natural selection—
History. I. Title.
QH361.G66 1980   575.01′62   80–15952

W. W. Norton & Company, Inc. 500 Fifth Avenue,
New York, N.Y. 10110
W. W. Norton & Company Ltd., 37 Great Russell Street,
London WC1B 3NU

7 8 9 0

ISBN 0-393-30023-4

Contents

The Panda's Thumb

More
Reflections
in Natural
History

Prologue

ON THE TITLE PAGE of his classic book, *The Cell in Development and Inheritance*, E.B. Wilson inscribed a motto from Pliny, the great natural historian who died in his boots when he sailed across the Bay of Naples to study the eruption of Mt. Vesuvius in A.D. 79. He suffocated in the same vapors that choked the citizens of Pompeii. Pliny wrote: *Natura nusquam magis est tota quam in minimis*—"Nature is to be found in her entirety nowhere more than in her smallest creatures." Wilson, of course, commandeered Pliny's statement to celebrate the microscopic building blocks of life, minute structures unknown perforce to the great Roman. Pliny was thinking about organisms.

Pliny's statement captures the essence of what fascinates me about natural history. In an old stereotype (not followed nearly so often as mythology proclaims), the natural history essay restricts itself to describing the peculiarities of animals—the mysterious ways of the beaver, or how the spider weaves her supple web. There is exultation in this and who shall gainsay it? But each organism can mean so much more to us. Each instructs; its form and behavior embodies general messages if only we can learn to read them. The language of this instruction is evolutionary theory. Exultation *and* explanation.

I was lucky to wander into evolutionary theory, one of the most exciting and important of all scientific fields. I had never heard of it when I started at a rather tender age; I was simply awed by dinosaurs. I thought paleontologists spent their lives digging up bones and putting them together, never venturing beyond the momentous issue of what connects to what. Then I discovered evolutionary theory. Ever since then, the duality of natural history—richness in particularities and potential union in underlying explanation—has propelled me.

I think that the fascination so many people feel for evolutionary theory resides in three of its properties. First, it is, in its current state of development, sufficiently firm to provide satisfaction and confidence, yet fruitfully undeveloped enough to provide a treasure trove of mysteries. Second, it stands in the middle of a continuum stretching from sciences that deal in timeless, quantitative generality to those that work directly with the singularities of history. Thus, it provides a home for all styles and propensities, from those who seek the purity of abstraction (the laws of population growth and the structure of DNA) to those who revel in the messiness of irreducible particularity (what, if anything, did *Tyrannosaurus* do with its puny front legs anyway?). Third, it touches all our lives; for how can we be indifferent to the great questions of genealogy: where did we come from, and what does it all mean? And then, of course, there are all those organisms: more than a million described species, from bacterium to blue whale, with one hell of a lot of beetles in between—each with its own beauty, and each with a story to tell.

These essays range broadly in the phenomena they treat —from the origin of life, to the brain of Georges Cuvier, to a mite that dies before it is born. Yet I hope that I have avoided that incubus of essay collections, diffuse incoherence, by centering them all upon evolutionary theory, with an emphasis on Darwin's thoughts and impact. As I stated in introducing my previous collection, *Ever Since Darwin:* "I am a tradesman, not a polymath. What I know of planets and politics lies at their intersection with biological evolution."

I have tried to weld these essays into an integrated whole

by organizing them into eight sections. The first on pandas, turtles & anglerfish, illustrates why we can be confident that evolution occurred. The argument embodies a paradox: the proof of evolution lies in imperfections that reveal history. This section is followed by a club sandwich—three sections on major themes in the evolutionary study of natural history (Darwinian theory and the meaning of adaptation, the tempo and mode of change, and the scaling of size and time), and two intervening layers of two sections each (III and IV, and VI and VII) on organisms and the peculiarities of their history. (If anyone wants to pursue the metaphorical sandwich and divide these seven sections into supporting structure and meat, I will not be offended.) I have also impaled the sandwich with toothpicks—subsidiary themes common to all sections, and intended to prick some conventional comforts: why science must be embedded in culture, why Darwinism cannot be squared with hopes for intrinsic harmony or progress in nature. But each pinprick has its positive consequence. An understanding of cultural bias forces us to view science as an accessible, human activity, much like any form of creativity. An abandonment of the hope that we might read a meaning for our lives passively in nature compels us to seek answers within ourselves.

These essays are lightly edited versions of my monthly columns in *Natural History* Magazine, collectively titled "This View of Life." I have added postscripts to a few: additional evidence of Teilhard's possible involvement in the Piltdown fraud (essay 10); a letter from J Harlen Bretz, controversial as ever at 96 (19); confirmation from the southern hemisphere for an explanation of magnets in bacteria (30). I thank Ed Barber for persuading me that these essays might be less ephemeral than I thought. *Natural History*'s editor in chief Alan Ternes and copy editor Florence Edelstein have greatly helped in deconvolution of phrase and thought and in devising some good titles. Four essays would not have been, without the gracious help of colleagues: Carolyn Fluehr-Lobban introduced me to Dr. Down, sent me his obscure article, and shared her insights and writing with me (essay 15). Ernst Mayr has urged the importance of folk taxonomy for years and had all the references on hand (essay 20). Jim Kennedy introduced me to Kirkpatrick's work (essay 22); otherwise I would never have penetrated the veil of silence surrounding it. Richard Frankel wrote me an unsolicited four-page letter explaining lucidly to this physical dunce the magnetic properties of his fascinating bacteria (essay 30). I am always cheered and delighted by the generosity of colleagues; a thousand untold stories overbalance every eagerly recorded case of nastiness. I thank Frank Sulloway for telling me the true story of Darwin's finches (essay 5), Diane Paul, Martha Denckla, Tim White, Andy Knoll, and Carl Wunsch for references, insights, and patient explanation.

Fortunately, I write these essays during an exciting time in evolutionary theory. When I think of paleontology in 1910, with its wealth of data and void of ideas, I regard it as a privilege to be working today.

Evolutionary theory is expanding its domain of impact and explanation in all directions. Consider the current excitement in such disparate realms as the basic mechanics of DNA, embryology, and the study of behavior. Molecular evolution is now a full-fledged discipline that promises to provide both strikingly new ideas (the theory of neutrality as an alternative to natural selection) and resolution of many classical mysteries in natural history (see essay 24). At the same time, the discovery of inserted sequences and jumping genes reveals a new stratum of genetic complexity that must be pregnant with evolutionary meaning. The triplet code is only a machine language; a higher level of control must exist. If we can ever figure out how multicellular creatures regulate the timing involved in the complex orchestration of their embryonic growth, then developmental biology might unite molecular genetics with natural history into a unified science of life. The theory of kin selection has extended Darwinian theory fruitfully into the realm of social behavior, though I believe that its more zealous advocates misunderstand the hierarchical nature of explanation and try to extend it (by more than permissible analogy) to realms of human culture where it does not apply (see essays 7 and 8).

Yet, while Darwinian theory extends its domain, some of its cherished postulates are slipping, or at least losing their generality. The "modern synthesis," the contemporary version of Darwinism that has reigned for thirty years, took the model of adaptive gene substitution within local populations as an adequate account, by accumulation and extension, of life's entire history. The model may work well in its empirical domain of minor, local, adaptive adjustment; populations of the moth *Biston betularia* did turn black, by substitution of a single gene, as a selected response for decreased visibility on trees that had been blackened by industrial soot. But is the origin of a new species simply this process extended to more genes and greater effect? Are larger evolutionary trends within major lineages just a further accumulation of sequential, adaptive changes?

Many evolutionists (myself included) are beginning to challenge this synthesis and to assert the hierarchical view that different levels of evolutionary change often reflect different kinds of causes. Minor adjustment within populations may be sequential and adaptive. But speciation may occur by major chromosomal changes that establish sterility with other species for reasons unrelated to adaptation. Evolutionary trends may represent a kind of higher-level selection upon essentially static species themselves, not the slow and steady alteration of a single large population through untold ages.

Before the modern synthesis, many biologists (see Bateson, 1922, in bibliography) expressed confusion and depression because the proposed mechanisms of evolution at different levels seemed contradictory enough to preclude a unified science. After the modern synthesis, the notion spread (amounting almost to a dogma among its less thoughtful lieutenants) that all evolution could be reduced to the basic Darwinism of gradual, adaptive change within local populations. I think that we are now pursuing a fruitful path between the anarchy of Bateson's day and the restriction of view imposed by the modern synthesis. The modern synthesis works in its appropriate arena, but the same Darwinian processes of mutation and selection may operate in strikingly different ways at higher domains in a hierarchy of evolutionary levels. I think that we may hope for uniformity of causal agents, hence a single, general theory with a Darwinian core. But we must reckon with a multiplicity of mechanisms that preclude the explanation of higher level phenomena by the model of adaptive gene substitution favored for the lowest level.

At the basis of all this ferment lies nature's irreducible complexity. Organisms are not billiard balls, propelled by simple and measurable external forces to predictable new positions on life's pool table. Sufficiently complex systems have greater richness. Organisms have a history that constrains their future in myriad, subtle ways (see essays of section I). Their complexity of form entails a host of functions incidental to whatever pressures of natural selection superintended the initial construction (see essay 4). Their intricate and largely unknown pathways of embryonic development guarantee that simple inputs (minor changes in timing, for example) may be translated into marked and surprising changes in output (the adult organism, see essay 18).

Charles Darwin chose to close his great book with a striking comparison that expresses this richness. He contrasted the simpler system of planetary motion, and its result of endless, static cycling, with the complexity of life and its wondrous and unpredictable change through the ages:

> There is grandeur in this view of life, with its several powers, having been originally breathed into a few forms or into one; and that, whilst this planet has gone cycling on according to the fixed law of gravity, from so simple a beginning endless forms most beautiful and most wonderful have been, and are being, evolved.

1 | Perfection and Imperfection: A Trilogy on a Panda's Thumb

1 | The Panda's Thumb

FEW HEROES LOWER their sights in the prime of their lives; triumph leads inexorably on, often to destruction. Alexander wept because he had no new worlds to conquer; Napoleon, overextended, sealed his doom in the depth of a Russian winter. But Charles Darwin did not follow the *Origin of Species* (1859) with a general defense of natural selection or with its evident extension to human evolution (he waited until 1871 to publish *The Descent of Man*). Instead, he wrote his most obscure work, a book entitled: *On the Various Contrivances by Which British and Foreign Orchids Are Fertilized by Insects* (1862).

Darwin's many excursions into the minutiae of natural history—he wrote a taxonomy of barnacles, a book on climbing plants, and a treatise on the formation of vegetable mold by earthworms—won him an undeserved reputation as an old-fashioned, somewhat doddering describer of curious plants and animals, a man who had one lucky insight at the right time. A rash of Darwinian scholarship has laid this myth firmly to rest during the past twenty years (see essay 2). Before then, one prominent scholar spoke for many ill-informed colleagues when he judged Darwin as a "poor joiner of ideas . . . a man who does not belong with the great thinkers."

In fact, each of Darwin's books played its part in the grand and coherent scheme of his life's work—demonstrating the fact of evolution and defending natural selection as its primary mechanism. Darwin did not study orchids solely for their own sake. Michael Ghiselin, a California biologist who finally took the trouble to read all of Darwin's books (see his *Triumph of the Darwinian Method*), has correctly identified the treatise on orchids as an important episode in Darwin's campaign for evolution.

Darwin begins his orchid book with an important evolutionary premise: continued self-fertilization is a poor strategy for long-term survival, since offspring carry only the genes of their single parent, and populations do not maintain enough variation for evolutionary flexibility in the face of environmental change. Thus, plants bearing flowers with both male and female parts usually evolve mechanisms to ensure cross-pollination. Orchids have formed an alliance with insects. They have evolved an astonishing variety of "contrivances" to attract insects, guarantee that sticky pollen adheres to their visitor, and ensure that the attached pollen comes in contact with female parts of the next orchid visited by the insect.

Darwin's book is a compendium of these contrivances, the botanical equivalent of a bestiary. And, like the medieval bestiaries, it is designed to instruct. The message is paradoxical but profound. Orchids manufacture their intricate devices from the common components of ordinary flowers, parts usually fitted for very different functions. If God had designed a beautiful machine to reflect his wisdom and power, surely he would not have used a collection of parts generally fashioned for other purposes. Orchids were not made by an ideal engineer; they are jury-rigged from a limited set of available components. Thus, they must have evolved from ordinary flowers.

Thus, the paradox, and the common theme of this trilogy of essays: Our textbooks like to illustrate evolution with examples of optimal design—nearly perfect mimicry of a dead leaf by a butterfly or of a poisonous species by a palatable relative. But ideal design is a lousy argument for evolution, for it mimics the postulated action of an omnipotent creator. Odd arrangements and funny solutions are the proof of evolution—paths that a sensible God would never tread but that a natural process, constrained by history, follows perforce. No one understood this better than Darwin. Ernst Mayr has shown how Darwin, in defending evolution, consistently turned to organic parts and geographic distributions that make the least sense. Which brings me to the giant panda and its "thumb."

Giant pandas are peculiar bears, members of the order Carnivora. Conventional bears are the most omnivorous representatives of their order, but pandas have restricted this catholicity of taste in the other direction—they belie the name of their order by subsisting almost entirely on bamboo. They live in dense forests of bamboo at high elevations in the mountains of western China. There they sit, largely unthreatened by predators, munching bamboo ten to twelve hours each day.

As a childhood fan of Andy Panda, and former owner of a stuffed toy won by some fluke when all the milk bottles actually tumbled at the county fair, I was delighted when the first fruits of our thaw with China went beyond ping pong to the shipment of two pandas to the Washington zoo. I went and watched in appropriate awe. They yawned, stretched, and ambled a bit, but they spent nearly all their time feeding on their beloved bamboo. They sat upright and manipulated the stalks with their forepaws, shedding the leaves and consuming only the shoots.

I was amazed by their dexterity and wondered how the scion of a stock adapted for running could use its hands so adroitly. They held the stalks of bamboo in their paws and stripped off the leaves by passing the stalks between an apparently flexible thumb and the remaining fingers. This puzzled me. I had learned that a dexterous, opposable thumb stood among the hallmarks of human success. We had maintained, even exaggerated, this important flexibility of our primate forebears, while most mammals had sacrificed it in specializing their digits. Carnivores run, stab, and scratch. My cat may manipulate me psychologically, but he'll never type or play the piano.

So I counted the panda's other digits and received an even greater surprise: there were five, not four. Was the "thumb" a separately evolved sixth finger? Fortunately, the giant panda has its bible, a monograph by D. Dwight Davis, late curator of vertebrate anatomy at Chicago's Field Museum of Natural History. It is probably the greatest work of modern evolutionary comparative anatomy, and it contains more than anyone would ever want to know about pandas. Davis had the answer, of course.

The panda's "thumb" is not, anatomically, a finger at all. It is constructed from a bone called the radial sesamoid, normally a small component of the wrist. In pandas, the radial sesamoid is greatly enlarged and elongated until it almost equals the metapodial bones of the true digits in length. The radial sesamoid underlies a pad on the panda's forepaw; the five digits form the framework of another pad, the palmar. A shallow furrow separates the two pads and serves as a channelway for bamboo stalks.

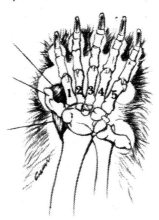

D. L. CRAMER

The panda's thumb comes equipped not only with a bone to give it strength but also with muscles to sustain its agility. These muscles, like the radial sesamoid bone itself, did not arise *de novo*. Like the parts of Darwin's orchids, they are familiar bits of anatomy remodeled for a new function. The abductor of the radial sesamoid (the muscle that pulls it away from the true digits) bears the formidable name *abduc-*

tor pollicis longus ("the long abductor of the thumb"—*pollicis* is the genitive of *pollex*, Latin for "thumb"). Its name is a giveaway. In other carnivores, this muscle attaches to the first digit, or true thumb. Two shorter muscles run between the radial sesamoid and the pollex. They pull the sesamoid "thumb" towards the true digits.

Does the anatomy of other carnivores give us any clue to the origin of this odd arrangement in pandas? Davis points out that ordinary bears and raccoons, the closest relatives of giant pandas, far surpass all other carnivores in using their forelegs for manipulating objects in feeding. Pardon the backward metaphor, but pandas, thanks to their ancestry, began with a leg up for evolving greater dexterity in feeding. Moreover, ordinary bears already have a slightly enlarged radial sesamoid.

In most carnivores, the same muscles that move the radial sesamoid in pandas attach exclusively to the base of the pollex, or true thumb. But in ordinary bears, the long abductor muscle ends in two tendons: one inserts into the base of the thumb as in most carnivores, but the other attaches to the radial sesamoid. The two shorter muscles also attach, in part, to the radial sesamoid in bears. "Thus," Davis concludes, "the musculature for operating this remarkable new mechanism—functionally a new digit—required no intrinsic change from conditions already present in the panda's closest relatives, the bears. Furthermore, it appears that the whole sequence of events in the musculature follows automatically from simple hypertrophy of the sesamoid bone."

The sesamoid thumb of pandas is a complex structure formed by marked enlargement of a bone and an extensive rearrangement of musculature. Yet Davis argues that the entire apparatus arose as a mechanical response to growth of the radial sesamoid itself. Muscles shifted because the enlarged bone blocked them short of their original sites. Moreover, Davis postulates that the enlarged radial sesamoid may have been fashioned by a simple genetic change, perhaps a single mutation affecting the timing and rate of growth.

In a panda's foot, the counterpart of the radial sesamoid, called the tibial sesamoid, is also enlarged, although not so much as the radial sesamoid. Yet the tibial sesamoid supports no new digit, and its increased size confers no advantage, so far as we know. Davis argues that the coordinated increase of both bones, in response to natural selection upon one alone, probably reflects a simple kind of genetic change. Repeated parts of the body are not fashioned by the action of individual genes—there is no gene "for" your thumb, another for your big toe, or a third for your pinky. Repeated parts are coordinated in development; selection for a change in one element causes a corresponding modification in others. It may be genetically more complex to enlarge a thumb and *not* to modify a big toe, than to increase both together. (In the first case, a general coordination must be broken, the thumb favored separately, and correlated increase of related structures suppressed. In the second, a single gene may increase the rate of growth in a field regulating the development of corresponding digits.)

The panda's thumb provides an elegant zoological counterpart to Darwin's orchids. An engineer's best solution is debarred by history. The panda's true thumb is committed to another role, too specialized for a different function to become an opposable, manipulating digit. So the panda must use parts on hand and settle for an enlarged wrist bone and a somewhat clumsy, but quite workable, solution. The sesamoid thumb wins no prize in an engineer's derby. It is, to use Michael Ghiselin's phrase, a contraption, not a lovely contrivance. But it does its job and excites our imagination all the more because it builds on such improbable foundations.

Darwin's orchid book is filled with similar illustrations. The marsh Epipactus, for example, uses its labellum—an enlarged petal—as a trap. The labellum is divided into two parts. One, near the flower's base, forms a large cup filled with nectar—the object of an insect's visit. The other, near the flower's edge, forms a sort of landing stage. An insect alighting on this runway depresses it and thus gains entrance to the nectar cup beyond. It enters the cup, but the

Marsh *Epipactis*, lower sepals removed

a. Runway of labellum depressed after insect lands.

D. L. CRAMER

b. Runway of labellum raised after insect crawls into cup below.

D. L. CRAMER

runway is so elastic that it instantly springs up, trapping the insect within the nectar cup. The insect must then back out through the only available exit—a path that forces it to brush against the pollen masses. A remarkable machine but all developed from a conventional petal, a part readily available in an orchid's ancestor.

Darwin then shows how the same labellum in other orchids evolves into a series of ingenious devices to ensure cross-fertilization. It may develop a complex fold that forces an insect to detour its proboscis around and past the pollen masses in order to reach nectar. It may contain deep channels or guiding ridges that lead insects both to nectar and pollen. The channels sometimes form a tunnel, producing a tubular flower. All these adaptations have been built from a part that began as a conventional petal in some ancestral form. Yet nature can do so much with so little that it displays, in Darwin's words, "a prodigality of resources for gaining the very same end, namely, the fertilization of one flower by pollen from another plant."

Darwin's metaphor for organic form reflects his sense of wonder that evolution can fashion such a world of diversity and adequate design with such limited raw material:

> Although an organ may not have been originally formed for some special purpose, if it now serves for this end we are justified in saying that it is specially contrived for it. On the same principle, if a man were to make a machine for some special purpose, but were to use old wheels, springs, and pulleys, only slightly altered, the whole machine, with all its parts, might be said to be specially contrived for that purpose. Thus throughout nature almost every part of each living being has probably served, in a slightly modified condition, for diverse purposes, and has acted in the living machinery of many ancient and distinct specific forms.

We may not be flattered by the metaphor of refurbished wheels and pulleys, but consider how well we work. Nature is, in biologist François Jacob's words, an excellent tinkerer, not a divine artificer. And who shall sit in judgment between these exemplary skills?

2 | Senseless Signs of History

WORDS PROVIDE CLUES about their history when etymology does *not* match current meaning. Thus, we suspect that emoluments were once fees paid to the local miller (from the Latin *molere,* to grind), while disasters must have been blamed upon evil stars.

Evolutionists have always viewed linguistic change as a fertile field for meaningful analogies. Charles Darwin, advocating an evolutionary interpretation for such vestigial structures as the human appendix and the embryonic teeth of whalebone whales, wrote: "Rudimentary organs may be compared with the letters in a word, still retained in the spelling, but become useless in the pronunciation, but which serve as a clue in seeking for its derivation." Both organisms and languages evolve.

This essay masquerades behind a list of curious facts, but it is really an abstract discourse on method—or, rather, on a particular method widely used but little appreciated by scientists. In the stereotyped image, scientists rely upon experiment and logic. A middle-aged man in a white coat (most stereotypes are sexist), either shyly reticent, but burning with an inner zeal for truth, or else ebullient and eccentric, pours two chemicals together and watches his answer emerge in a flask. Hypotheses, predictions, experiments, and answers: the scientific method.

But many sciences do not and cannot work this way. As a paleontologist and evolutionary biologist, my trade is the reconstruction of history. History is unique and complex. It cannot be reproduced in a flask. Scientists who study history, particularly an ancient and unobservable history not recorded in human or geological chronicles, must use inferential rather than experimental methods. They must examine *modern results* of historical processes and try to reconstruct the path leading from ancestral to contemporary words, organisms, or landforms. Once the path is traced, we may be able to specify the causes that led history to follow this, rather than another, route. But how can we infer pathways from modern results? In particular, how can we be sure that there was a pathway at all? How do we know that a modern result is the product of alteration through history and not an immutable part of a changeless universe?

This is the problem that Darwin faced, for his creationist opponents did view each species as unaltered from its initial formation. How did Darwin prove that modern species are the products of history? We might suppose that he looked toward the most impressive results of evolution, the complex and perfected adaptations of organisms to their environments: the butterfly passing for a dead leaf, the bittern for a branch, the superb engineering of a gull aloft or a tuna in the sea.

Paradoxically, he did just the opposite. He searched for oddities and imperfections. The gull may be a marvel of design; if one believes in evolution beforehand, then the engineering of its wing reflects the shaping power of natural selection. But you cannot demonstrate evolution with perfection because perfection need not have a history. After all, perfection of organic design had long been the favorite argument of creationists, who saw in consummate engineering the direct hand of a divine architect. A bird's wing, as an aerodynamic marvel, might have been created exactly as we find it today.

But, Darwin reasoned, if organisms have a history, then ancestral stages should leave *remnants* behind. Remnants of the past that don't make sense in present terms—the useless, the odd, the peculiar, the incongruous—are the signs of history. They supply proof that the world was not made in its present form. When history perfects, it covers its own tracks.

Why should a general word for monetary compensation refer literally to a profession now virtually extinct, unless it once had some relation with grinding and grain? And why should the fetus of a whale make teeth in its mother's womb only to resorb them later and live a life sifting krill on a whalebone filter, unless its ancestors had functional teeth and these teeth survive as a remnant during a stage when .they do no harm?

No evidence for evolution pleased Darwin more than the presence in nearly all organisms of rudimentary or vestigial structures, "parts in this strange condition, bearing the stamp of unutility," as he put it. "On my view of descent with modification, the origin of rudimentary organs is simple," he continued. They are bits of useless anatomy, preserved as remnants of functional parts in ancestors.

The general point extends both beyond rudimentary structures and beyond biology to any historical science. Oddities in current terms are the signs of history. The first essay of this trilogy raised the same subject in a different context. The panda's "thumb" demonstrates evolution *because* it is clumsy and built from an odd part, the radial sesamoid bone of the wrist. The true thumb had been so shaped in its ancestral role as the running and clawing digit of a carnivore that it could not be modified into an opposable grasper for bamboo in a vegetarian descendant.

In a nonbiological musing, I found myself wondering last week why *veteran* and *veterinarian,* two words with such different meanings, should have a similar root in the Latin *vetus,* or "old." Again, an oddity suggesting a genealogical approach for its solution. Veteran presented no problem, for its root and its modern meaning coincide—no indication of history. Veterinarian turned out to be interesting. City dwellers tend to view vets as servants of their pampered dogs and cats. I forgot that the original veterinarians treated farm and herd animals (as do most modern vets, I guess—pardon my New Yorker's parochialism). The link to *vetus* is through "beast of burden"—old, in the sense of "able to take a load." Cattle, in Latin, are *veterinae*.

This general principle of historical science should apply to the earth as well. The theory of plate tectonics has led us to reconstruct the history of our planet's surface. During the past 200 million years, our modern continents have fragmented and dispersed from a single supercontinent, Pangaea, that coalesced from earlier continents more than 225 million years ago. If modern oddities are the signs of history, we should ask whether any peculiar things that animals do today might be rendered more sensible as adaptations to previous continental positions. Among the greatest puzzles and wonders of natural history are the long and circuitous routes of migration followed by many animals. Some lengthy movements make sense as direct paths to favorable climates from season to season; they are no more peculiar than the annual winter migration to Florida of large mammals inside metallic birds. But other animals migrate thousands of miles—from feeding to breeding grounds—with astounding precision when other appropriate spots seem close at hand. Could any of these peculiar routes be rendered shorter and more sensible on a map of ancient continental positions? Archie Carr, world's expert on the migration of green turtles, has made such a proposal.

A population of the green turtle, *Chelonia mydas*, nests and breeds on the small and isolated central Atlantic island of Ascension. London soup chefs and victualing ships of Her Majesty's Navy found and exploited these turtles long ago. But they did not suspect, as Carr discovered by tagging animals at Ascension and recovering them later at their feeding grounds, that *Chelonia* travels 2,000 miles from the coast of Brazil to breed on this "pinpoint of land hundreds of miles from other shores," this "barely exposed spire in mid-ocean."

Turtles feed and breed on separate grounds for good reasons. They feed on sea grasses in protected, shallow-water pastures, but breed on exposed shores where sandy beaches develop—preferably, on islands where predators are rare. But why travel 2,000 miles to the middle of an ocean when other, apparently appropriate breeding grounds are so much nearer? (Another large population of the same species breeds on the Caribbean coast of Costa Rica.) As Carr writes: "The difficulties facing such a voyage would seem insurmountable if it were not so clear that the turtles are somehow surmounting them."

Perhaps, Carr reasoned, this odyssey is a peculiar extension of something much more sensible, a journey to an island in the middle of the Atlantic, when the Atlantic was little more than a puddle between two continents recently separated. South America and Africa parted company some 80 million years ago, when ancestors of the genus *Chelonia* were already present in the area. Ascension is an island associated with the Mid-Atlantic Ridge, a linear belt where new sea floor wells up from the earth's interior. This upwelling material often piles itself high enough to form islands.

Iceland is the largest modern island formed by the Mid-Atlantic Ridge; Ascension is a smaller version of the same process. After islands form on one side of a ridge, they are pushed away by new material welling up and spreading out. Thus, islands tend to be older as we move farther and farther from a ridge. But they also tend to get smaller and finally to erode away into underwater seamounts, for their supply of new material dries up once they drift away from an active ridge. Unless preserved and built up by a shield of coral or other organisms, islands will eventually be eroded below sea level by waves. (They may also sink gradually from sight as they move downslope from an elevated ridge into the oceanic depths.)

Carr therefore proposed that the ancestors of Ascension green turtles swam a short distance from Brazil to a "proto-Ascension" on the late Cretaceous Mid-Atlantic Ridge. As this island moved out and sank, a new one formed at the ridge and the turtles ventured a bit farther. This process continued until, like the jogger who does a bit more each day and ends up a marathoner, turtles found themselves locked into a 2,000-mile journey. (This historical hypothesis does not deal with the other fascinating question of how the turtles can find this dot in a sea of blue. The hatchlings float to Brazil on the Equatorial Current, but how do they get back? Carr supposes that they begin their journey by celestial cues and finally home in by remembering the character [taste? smell?] of Ascension water when they detect the island's wake.)

Carr's hypothesis is an excellent example of using the peculiar to reconstruct history. I wish I could believe it. I am not troubled by the empirical difficulties, for these do not render the theory implausible. Can we be confident, for example, that a new island always arose in time to replace an old one—for the absence of an island for even one generation would disrupt the system. And would the new islands always arise sufficiently "on course" to be found? Ascension itself is less than seven million years old.

I am more bothered by a theoretical difficulty. If the entire species *Chelonia mydas* migrated to Ascension or, even better, if a group of related species made the journey, I would have no objection, for behavior can be as ancient and as heritable as form. But *C. mydas* lives and breeds throughout the world. The Ascension turtles represent only one among many breeding populations. Although its ancient ancestors may have lived in the Atlantic puddle 200 million years ago, our record of the genus *Chelonia* does not extend back beyond fifteen million years, while the species *C. mydas* is probably a good deal younger. (The fossil record, for all its faults, indicates that few vertebrate species survive for as many as ten million years.) In Carr's scheme, the turtles that made the first trips to proto-Ascension were rather distant ancestors of *C. mydas* (in a different genus at least). Several events of speciation separate this Cretaceous ancestor from the modern green turtle. Now consider what must have happened if Carr is right. The ancestral species must have been divided into several breeding populations, only one of which went to proto-Ascension. This species then evolved to another and another through however many evolutionary steps separated it from *C. mydas*. At each step, the Ascension population kept its integrity, changing in lock step with other separate populations from species to species.

But evolution, so far as we know, doesn't work this way.

New species arise in small, isolated populations and spread out later. Separate subpopulations of a widely dispersed species do not evolve in parallel from one species to the next. If the subpopulations are separate breeding stocks, what is the chance that all would evolve in the same way and still be able to interbreed when they had changed enough to be called a new species? I assume that *C. mydas,* like most species, arose in a small area sometime within the last ten million years, when Africa and South America were not much closer together than they are today.

In 1965, before continental drift became fashionable, Carr proposed a different explanation that makes more sense to me because it derives the Ascension population after *C. mydas* evolved. He argued that ancestors of the Ascension population accidentally drifted on the Equatorial Current from west Africa to Ascension. (Carr points out that another turtle, the west African ridley, *Lepidochelys olivacea,* colonized the South American coast by this route.) The hatchlings then drifted to Brazil in the same east-to-west current. Of course, getting back to Ascension is the problem, but the mechanism of turtle migration is so mysterious that I see no barrier to supposing that turtles can be imprinted to remember the place of their birth without prior genetic information transmitted from previous generations.

I don't think that the validation of continental drift is the only factor that caused Carr to change his mind. He implies that he favors his new theory because it preserves some basic styles of explanation generally preferred by scientists (incorrectly, in my iconoclastic opinion). By Carr's new theory, the peculiar Ascension route evolved gradually, in a sensible and predictable fashion, step by step. In his former view, it is a sudden event, an accidental, unpredictable vagary of history. Evolutionists tend to be more comfortable with nonrandom, gradualistic theories. I think that this is a deep prejudice of Western philosophical traditions, not a reflection of nature's ways (see essays of section 5). I regard Carr's new theory as a daring hypothesis in support of a conventional philosophy. I suspect that it is wrong, but I applaud his ingenuity, his effort, and his method, for he follows the great historical principle of using the peculiar as a sign of change.

I am afraid that the turtles illustrate another aspect of historical science—this time a frustration, rather than a principle of explanation. Results rarely specify their causes unambiguously. If we have no direct evidence of fossils or human chronicles, if we are forced to infer a process only from its modern results, then we are usually stymied or reduced to speculation about probabilities. For many roads lead to almost any Rome.

This round goes to the turtles—and why not? While Portuguese sailors hugged the coast of Africa, *Chelonia mydas* swam straight for a dot in the ocean. While the world's best scientists struggled for centuries to invent the tools of navigation, *Chelonia* looked at the skies and proceeded on course.

3 | Double Trouble

NATURE MARKS Izaak Walton as a rank amateur more often than I had imagined. In 1654, the world's most famous fisherman before Ted Williams wrote of his favorite lure: "I have an artificial minnow . . . so curiously wrought, and so exactly dissembled that it would beguile any sharpsighted trout in a swift stream."

An essay in my previous book, *Ever Since Darwin,* told the tale of *Lampsilis,* a freshwater clam with a decoy "fish" mounted on its rear end. This remarkable lure has a streamlined "body," side flaps simulating fins and tail, and an eyespot for added effect; the flaps even undulate with a rhythmic motion that imitates swimming. This "fish," constructed from a brood pouch (the body) and the clam's outer skin (fin and tails), attracts the real item and permits a mother clam to shoot her larvae from the brood pouch toward an unsuspecting fish. Since the larvae of *Lampsilis* can only grow as parasites on a fish's gill, this decoy is a useful device indeed.

I was astounded recently to learn that *Lampsilis* is not alone. Ichthyologists Ted Pietsch and David Grobecker recovered a single specimen of an amazing Philippine anglerfish, not as a reward for intrepid adventures in the wilds, but from that source of so much scientific novelty—the local aquarium retailer. (Recognition, rather than *machismo,* is often the basis of exotic discovery.) Anglerfish lure their dinner, rather than a free ride for their larvae. They carry a highly modified dorsal fin spine affixed to the tips of their snouts. At the end of this spine, they mount an appropriate lure. Some deep-sea species, living in a dark world untouched by light from the surface, fish with their own source of illumination: they gather phosphorescent bacteria in their lures. Shallow-water species tend to have colorful, bumpy bodies, and look remarkably like rocks encrusted with sponges and algae. They rest inert on the bottom and wave or wiggle their conspicuous lures near their mouths. "Baits" differ among species, but most resemble—often imperfectly—a variety of invertebrates, including worms and crustaceans.

Anglerfish   DAVID B. GROBECKER

Pietsch and Grobecker's anglerfish, however, has evolved a fish lure every bit as impressive as the decoy mounted on *Lampsilis*'s rear—a first for anglerfish. (Their report bears as its appropriate title "The Compleat Angler" and cites as an epigraph the passage from Walton quoted above.) This exquisite fake also sports eyelike spots of pigment in the right place. In addition, it bears compressed filaments representing pectoral and pelvic fins along the bottom of the body, extensions from the back resembling dorsal and anal fins, and even an expanded rear projection looking for all the world like a tail. Pietsch and Grobecker conclude: "The bait is nearly an exact replica of a small fish that could easily belong to any of a number of percoid families common to the Philippine region." The angler even ripples its bait through the water, "simulating the lateral undulations of a swimming fish."

These nearly identical artifices of fish and clam might seem, at first glance, to seal the case for Darwinian evolution. If natural selection can do this twice, surely it can do anything. Yet—continuing the theme of the last two essays and bringing this trilogy to a close—perfection works as well for the creationist as the evolutionist. Did not the psalmist proclaim: "The heavens declare the glory of God; and the firmament showeth his handiwork." The last two essays argued that imperfection carries the day for evolution. This one discusses the Darwinian response to perfection.

The only thing more difficult to explain than perfection is repeated perfection by very different animals. A fish on a clam's rear end *and* another in front of an anglerfish's nose—the first evolved from a brood pouch and outer skin, the second from a fin spine—more than doubles the trouble. I have no difficulty defending the origin of both "fishes" by evolution. A plausible series of intermediate stages can be identified for *Lampsilis*. The fact that anglerfish press a fin spine into service as a lure reflects the jury-rigged, parts-available principle that made the panda's thumb and the orchid's labellum speak so strongly for evolution (see the first essay of this trilogy). But Darwinians must do more than demonstrate evolution; they must defend the basic mechanism of random variation and natural selection as the primary cause of evolutionary change.

Anti-Darwinian evolutionists have always favored the *re-*

peated development of very similar adaptations in different lineages as an argument against the central Darwinian notion that evolution is unplanned and undirected. If different organisms converge upon the same solutions again and again, does this not indicate that certain directions of change are preset, not established by natural selection working on random variation? Should we not look upon the repeated form itself as a cause of the numerous evolutionary events leading toward it?

Throughout his last half-dozen books, for example, Arthur Koestler has been conducting a campaign against his own misunderstanding of Darwinism. He hopes to find some ordering force, constraining evolution to certain directions and overriding the influence of natural selection. Repeated evolution of excellent design in separate lineages is his bulwark. Again and again, he cites the "nearly identical skulls" of wolves and the "Tasmanian wolf." (This marsupial carnivore looks like a wolf but is, by genealogy, more closely related to wombats, kangaroos, and koalas.) In *Janus,* his latest book, Koestler writes: "Even the evolution of a single species of wolf by random mutation plus selection presents, as we have seen, insurmountable difficulties. To duplicate this process independently on island and mainland would mean squaring a miracle."

The Darwinian response involves both a denial and an explanation. First, the denial: it is emphatically not true that highly convergent forms are effectively identical. Louis Dollo, the great Belgian paleontologist who died in 1931, established a much misunderstood principle—"the irreversibility of evolution" (also known as Dollo's law). Some ill-informed scientists think that Dollo advocated a mysterious directing force, driving evolution forward, never permitting a backward peek. And they rank him among the non-Darwinians who feel that natural selection cannot be the cause of nature's order.

In fact, Dollo was a Darwinian interested in the subject of convergent evolution—the repeated development of similar adaptations in different lineages. Elementary probability theory, he argued, virtually guarantees that convergence can never yield anything close to perfect resemblance. Organisms cannot erase their past. Two lineages may develop remarkable, superficial similarities as adaptations to a common mode of life. But organisms contain so many complex and independent parts that the chance of all evolving twice toward exactly the same result is effectively nil. Evolution is irreversible; signs of ancestry are always preserved; convergence, however impressive, is always superficial.

Consider my candidate for the most astounding convergence of all: the ichthyosaur. This sea-going reptile with terrestrial ancestors converged so strongly on fishes that it actually evolved a dorsal fin and tail in just the right place and with just the right hydrological design. These structures are all the more remarkable because they evolved from nothing—the ancestral terrestrial reptile had no hump on its back or blade on its tail to serve as a precursor. Nonetheless, the ichthyosaur is no fish, either in general design or in intricate detail. (In ichthyosaurs, for example, the vertebral column runs through the lower tail blade; in fish with tail vertebrae, the column runs into the upper blade.) The ichthyosaur remains a reptile, from its lungs and surface breathing to its flippers made of modified leg bones, not fin rays.

Ichthyosaur
COURTESY OF THE AMERICAN MUSEUM OF NATURAL HISTORY

Koestler's carnivores tell the same tale. Both placental wolf and marsupial "wolf" are well designed to hunt, but no expert would ever mistake their skulls. The numerous, small marks of marsupiality are not obliterated by convergence in outward form and function.

Second, the explanation: Darwinism is not the theory of capricious change that Koestler imagines. Random variation may be the raw material of change, but natural selection builds good design by rejecting most variants while accepting and accumulating the few that improve adaptation to local environments.

The basic reason for strong convergence, prosaic though it may seem, is simply that some ways of making a living impose exacting criteria of form and function upon any organism playing the role. Mammalian carnivores must run and stab; they do not need grinding molar teeth since they tear and swallow their food. Both placental and marsupial wolves are built for sustained running, have long, sharp, pointed canine teeth and reduced molars. Terrestrial vertebrates propel themselves with their limbs and may use their tails for balance. Swimming fish balance with their fins and propel from the rear with their tails. Ichthyosaurs, living like fish, evolved a broad propulsive tail (as whales did later —although the horizontal flukes of a whale's tail beat up and down, while the vertical flukes of fish and ichthyosaurs beat from side to side).

No one has treated this biological theme of repeated, exquisite design more eloquently than D'Arcy Wentworth Thompson in his 1942 treatise, *On Growth and Form*, still in print and still as relevant as ever. Sir Peter Medawar, a man who eschews hype and exaggeration, describes it as "beyond comparison the finest work of literature in all the annals of science that have been recorded in the English tongue." Thompson, zoologist, mathematician, classical scholar, and prose stylist, won accolades as an old man but spent his entire professional life in a small Scottish university because his views were too unorthodox to win prestigious London and Oxbridge jobs.

Thompson was more a brilliant reactionary than a visionary. He took Pythagoras seriously and worked as a Greek geometrician. He took special delight in finding the abstract forms of an idealized world embodied again and again in the products of nature. Why do repeated hexagons appear in the cells of a honeycomb and in the interlocking plates of some turtle shells? Why do the spirals in a pine cone and a sunflower (and often of leaves on a stem) follow the Fibonacci series? (A system of spirals radiating from a common point can be viewed either as a set of left- or right-handed spirals. Left and right spirals are not equal in number, but represent two consecutive figures of the Fibonacci series. The Fibonacci series is constructed by adding the previous two numbers to form the next: 1, 1, 2, 3, 5, 8, 13, 21, etc. The pine cone may, for example, have 13 left spirals and 21 right spirals.) Why do so many snail shells, ram's horns, and even the path of a moth to light follow a curve called the logarithmic spiral?

Thompson's answer was the same in each case: these abstract forms are optimal solutions for common problems. They are evolved repeatedly in disparate groups because they are the best, often the only, path to adaptation. Triangles, parallelograms, and hexagons are the only plane figures that fill space completely without leaving holes. Hexagons are often favored because they approximate a circle and maximize area within relative to the supporting walls (minimum construction for greatest storage of honey, for example). The Fibonacci pattern emerges automatically in any system of radiating spirals built by adding new elements at the apex, one at a time in the largest space available. The logarithmic spiral is the only curve that does not change its shape as it grows in size. I can identify the abstract Thompsonian forms as optimal adaptations, but to the larger metaphysical issue of why "good" form often exhibits such simple, numerical regularity, I plead only ignorance and wonder.

So far, I have only spoken to half the issue embodied in the problem of repeated perfection. I have discoursed on the "why." I have argued that convergence never renders two complex organisms completely identical (a circumstance that would strain Darwinian processes beyond their reasonable power) and I have tried to explain close repeats as optimal adaptations to common problems with few solutions.

But what about the "how?" We may know what the fish

of *Lampsilis* and the lure of the anglerfish are for, but how did they arise? This problem becomes particularly acute when the final adaptation is complex and peculiar but built from familiar parts of different ancestral function. If the angler's fishlike lure required 500 entirely separate modifications to attain its exquisite mimicry, then how did the process begin? And why did it continue, unless some non-Darwinian force, cognizant of the final goal, drove it on? Of what possible benefit is step one alone? Is a five-hundredth of a fake enough to inspire the curiosity of any real item?

D'Arcy Thompson's answer to this problem was overextended but characteristically prophetic. He argued that organisms are shaped directly by physical forces acting upon them: optima of form are nothing more than the natural states of plastic matter in the presence of appropriate physical forces. Organisms jump suddenly from one optimum to another when the regime of physical forces alters. We now know that physical forces are too weak, in most cases, to build form directly—and we look to natural selection instead. But we are derailed if selection can only act in a patient and piecemeal way—step by sequential step to build any complex adaptation.

I believe that a solution lies in the essence of Thompson's insight, shorn of his unsubstantiated claim that physical forces shape organisms directly. Complex forms are often built by a much simpler (often a very simple) system of generating factors. Parts are connected in intricate ways through growth, and alteration of one may resound through the entire organism and change it in a variety of unsuspected ways. David Raup, of Chicago's Field Museum of Natural History, adapted D'Arcy Thompson's insight to a modern computer, and showed that the basic forms of coiled shells—from nautiloid to clam to snail—can all be generated by varying only three simple gradients of growth. Using Raup's program, I can change a garden-variety snail into a common clam by modifying just two of the three gradients. And, believe it or not, a peculiar genus of modern snails does carry a bivalved shell so like a conventional

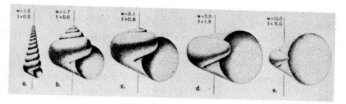

In these computer-drawn figures (they are not real snails, despite the similarities), a form (right) looking much like certain clams can be converted into a "snail" (left figures) simply by decreasing the rate at which the generating ellipse increases as the "shell" grows and by increasing the rate of translation of this ellipse down the axis of coiling. All these figures are drawn by specifying just four parameters. PHOTO COURTESY OF D. M. RAUP

clam's that I gasped when I saw a snail's head poking out between the valves in a striking close-up movie.

This closes my trilogy on the issue of perfection and imperfection as signs of evolution. But the entire set is really an extended disquisition on the panda's "thumb," a single, concrete object that spawned all three essays, despite their subsequent wanderings and musings. The thumb, built of a wrist bone, imperfect as a sign of history, constructed from parts available. Dwight Davis faced the dilemma of potential impotence for natural selection if it must work step by countless step to make a panda from a bear. And he advocated D'Arcy Thompson's solution of reduction to a simple system of generating factors. He showed how the complex apparatus of the thumb, with all its muscles and nerves, may arise as a set of automatic consequences following a simple enlargement of the radial sesamoid bone. He then argued that the complex changes in form and function of the skull—the transition from omnivory to nearly exclusive munching on bamboo—could be expressed as consequences of one or two underlying modifications. He concluded that "very few genetic mechanisms—perhaps no more than half a dozen—were involved in the primary adaptive shift from *Ursus* [bear] to

Ailuropoda [panda]. The action of most of these mechanisms can be identified with reasonable certainty."

And thus we may pass from the underlying genetic continuity of change—an essential Darwinian postulate—to a potentially episodic alteration in its manifest result—a sequence of complex, adult organisms. Within complex systems, smoothness of input can translate into episodic change in output. Here we encounter a central paradox of our being and of our quest to understand what made us. Without this level of complexity in construction, we could not have evolved the brains to ask such questions. With this complexity, we cannot hope to find solutions in the simple answers that our brains like to devise.

2 | Darwiniana

4 | Natural Selection and the Human Brain: Darwin *vs.* Wallace

IN THE SOUTH transept of Chartres cathedral, the most stunning of all medieval windows depicts the four evangelists as dwarfs sitting upon the shoulders of four Old Testament prophets—Isaiah, Jeremiah, Ezekiel, and Daniel. When I first saw this window as a cocky undergraduate in 1961, I immediately thought of Newton's famous aphorism—"if I have seen farther, it is by standing on the shoulders of giants"—and imagined that I had made a major discovery in unearthing his lack of originality. Years later, and properly humbled for many reasons, I learned that Robert K. Merton, the celebrated sociologist of science from Columbia University, had devoted an entire book to pre-Newtonian usages of the metaphor. It is titled, appropriately, *On The Shoulders of Giants.* In fact, Merton traces the bon mot back to Bernard of Chartres in 1126 and cites several scholars who believe that the windows of the great south transept, installed after Bernard's death, represent an explicit attempt to capture his metaphor in glass.

Although Merton wisely constructs his book as a delightful romp through the intellectual life of medieval and Renaissance Europe, he does have a serious point to make. For Merton has devoted much of his work to the study of multiple discoveries in science. He has shown that almost all major ideas arise more than once, independently and often virtually at the same time—and thus, that great scientists are embedded in their cultures, not divorced from them. Most great ideas are "in the air," and several scholars simultaneously wave their nets.

One of the most famous of Merton's "multiples" resides in my own field of evolutionary biology. Darwin, to recount the famous tale briefly, developed his theory of natural selection in 1838 and set it forth in two unpublished sketches of 1842 and 1844. Then, never doubting his theory for a moment, but afraid to expose its revolutionary implications, he proceeded to stew, dither, wait, ponder, and collect data for another fifteen years. Finally, at the virtual insistence of his closest friends, he began to work over his notes, intending to publish a massive tome that would have been four times as long as the *Origin of Species*. But, in 1858, Darwin received a letter and manuscript from a young naturalist, Alfred Russel Wallace, who had independently constructed the theory of natural selection while lying ill with malaria on an island in the Malay Archipelago. Darwin was stunned by the detailed similarity. Wallace even claimed inspiration from the same nonbiological source—Malthus' *Essay on Population*. Darwin, in great anxiety, made the expected gesture of magnanimity, but devoutly hoped that some way might be found to preserve his legitimate priority. He wrote to Lyell: "I would far rather burn my whole book, than that he or any other man should think that I have behaved in a paltry spirit." But he added a suggestion: "If I could honorably publish, I would state that I was induced now to publish a sketch . . . from Wallace having sent me an outline of my general conclusions." Lyell and Hooker took the bait and came to Darwin's rescue. While Darwin stayed home, mourning the death of his young child from scarlet fever, they presented a joint paper to the Linnaean Society containing an excerpt from Darwin's 1844 essay together with Wallace's manuscript. A year later, Darwin published his feverishly compiled "abstract" of the longer work—the *Origin of Species*. Wallace had been eclipsed.

Wallace has come down through history as Darwin's shadow. In public and private, Darwin was infallibly decent and generous to his younger colleague. He wrote to Wallace in 1870: "I hope it is a satisfaction to you to reflect—

and very few things in my life have been more satisfactory to me—that we have never felt any jealousy towards each other, though in one sense rivals." Wallace, in return, was consistently deferential. In 1864, he wrote to Darwin: "As to the theory of Natural Selection itself, I shall always maintain it to be actually yours and yours only. You had worked it out in details I had never thought of, years before I had a ray of light on the subject, and my paper would never have convinced anybody or been noticed as more than an ingenious speculation, whereas your book has revolutionized the study of Natural History, and carried away captive the best men of the present age."

This genuine affection and mutual support masked a serious disagreement on what may be the fundamental question in evolutionary theory—both then and today. How exclusive is natural selection as an agent of evolutionary change? Must all features of organisms be viewed as adaptations? Yet Wallace's role as Darwin's subordinate alter ego is so firmly fixed in popular accounts that few students of evolution are even aware that they ever differed on theoretical questions. Moreover, in the one specific area where their public disagreement is a matter of record—the origin of human intellect—many writers have told the story backwards because they failed to locate this debate in the context of a more general disagreement on the power of natural selection.

All subtle ideas can be trivialized, even vulgarized, by portrayal in uncompromising and absolute terms. Marx felt compelled to deny that he was a marxist, while Einstein contended with the serious misstatement that he meant to say "all is relative." Darwin lived to see his name appropriated for an extreme view that he never held—for "Darwinism" has often been defined, both in his day and in our own, as the belief that virtually all evolutionary change is the product of natural selection. In fact Darwin often complained, with uncharacteristic bitterness, about this misappropriation of his name. He wrote in the last edition of the *Origin* (1872): "As my conclusions have lately been much misrepresented, and it has been stated that I attribute the modification of species exclusively to natural selection, I may be permitted to remark that in the first edition of this work, and subsequently, I placed in a most conspicuous position—namely, at the close of the Introduction—the following words: 'I am convinced that natural selection has been the main but not the exclusive means of modification.' This has been of no avail. Great is the power of steady misrepresentation.''

However, England did house a small group of strict selectionists—"Darwinians" in the misappropriated sense—and Alfred Russel Wallace was their leader. These biologists did attribute all evolutionary change to natural selection. They viewed each bit of morphology, each function of an organ, each behavior as an adaptation, a product of selection leading to a "better" organism. They held a deep belief in nature's "rightness," in the exquisite fit of all creatures to their environments. In a curious sense, they almost reintroduced the creationist notion of natural harmony by substituting an omnipotent force of natural selection for a benevolent deity. Darwin, on the other hand, was a consistent pluralist gazing upon a messier universe. He saw much fit and harmony, for he believed that natural selection holds pride of place among evolutionary forces. But other processes work as well, and organisms display an array of features that are not adaptations and do not promote survival directly. Darwin emphasized two principles leading to nonadaptive change: (1) organisms are integrated systems and adaptive change in one part can lead to nonadaptive modifications of other features ("correlations of growth" in Darwin's phrase); (2) an organ built under the influence of selection for a specific role may be able, as a consequence of its structure, to perform many other, unselected functions as well.

Wallace stated the hard hyper-selectionist line—"pure Darwinism" in his terms—in an early article of 1867, calling it "a necessary deduction from the theory of natural selection."

None of the definite facts of organic selection, no special organ, no characteristic form or marking, no peculiarities of instinct or of habit, no relations between species or between groups of species, can exist but which must now be, or once have been, useful to the individuals or races which possess them.

Indeed, he argued later, any apparent nonutility must only reflect our faulty knowledge—a remarkable argument since it renders the principle of utility impervious to disproof a priori: "The assertion of 'inutility' in the case of any organ . . . is not, and can never be, the statement of a fact, but merely an expression of our ignorance of its purpose or origin."

All the public and private arguments that Darwin pursued with Wallace centered upon their differing assessments of the power of natural selection. They first crossed swords on the issue of "sexual selection," the subsidiary process that Darwin had proposed in order to explain the origin of features that appeared to be irrelevant or even harmful in the usual "struggle for existence" (expressed primarily in feeding and defense), but that could be interpreted as devices for increasing success in mating—elaborate antlers of deer, or tail feathers of the peacock, for example. Darwin proposed two kinds of sexual selection—competition among males for access to females, and choice exercised by females themselves. He attributed much of the racial differentiation among modern humans to sexual selection, based upon different criteria of beauty that arose among various peoples. (His book on human evolution—*The Descent of Man* (1871)—is really an amalgam of two works: a long treatise on sexual selection throughout the animal kingdom, and a shorter speculative account of human origins, relying heavily upon sexual selection.)

The notion of sexual selection is not really contrary to natural selection, for it is just another route to the Darwinian imperative of differential reproductive success. But Wallace disliked sexual selection for three reasons: it compromised the generality of that peculiarly nineteenth-century view of natural selection as a battle for life itself, not merely for copulation; it placed altogether too much emphasis upon the "volition" of animals, particularly in the concept of female choice; and, most importantly, it permitted the development of numerous, important features that are irrelevant, if not actually harmful, to the operation of an organism as a well-designed machine. Thus, Wallace viewed sexual selection as a threat to his vision of animals as works of exquisite craftsmanship, wrought by the purely material force of natural selection. (Indeed, Darwin had developed the concept largely to explain why so many differences among human groups are irrelevant to survival based upon good design, but merely reflect the variety of capricious criteria for beauty that arose for no adaptive reason among various races. Wallace did accept sexual selection based upon male combat as close enough to the metaphor of battle that controlled his concept of natural selection. But he rejected the notion of female choice, and greatly distressed Darwin with his speculative attempts to attribute all features arising from it to the adaptive action of natural selection.)

In 1870, as he prepared the *Descent of Man,* Darwin wrote to Wallace: "I grieve to differ from you, and it actually terrifies me and makes me constantly distrust myself. I fear we shall never quite understand each other." He struggled to understand Wallace's reluctance and even to accept his friend's faith in unalloyed natural selection: "You will be pleased to hear," he wrote to Wallace, "that I am undergoing severe distress about protection and sexual selection; this morning I oscillated with joy towards you; this evening I have swung back to [my] old position, out of which I fear I shall never get."

But the debate on sexual selection was merely a prelude to a much more serious and famous disagreement on that most emotional and contentious subject of all—human origins. In short, Wallace, the hyper-selectionist, the man who had twitted Darwin for his unwillingness to see the action of natural selection in every nuance of organic form, halted abruptly before the human brain. Our intellect and morality, Wallace argued, could not be the product of natural selection; therefore, since natural selection is evolution's only way, some higher power—God, to put it directly—must have intervened to construct this latest and greatest of organic innovations.

If Darwin had been distressed by his failure to impress Wallace with sexual selection, he was now positively aghast at Wallace's abrupt about-face at the finish line itself. He wrote to Wallace in 1869: "I hope you have not murdered too completely your own and my child." A month later, he remonstrated: "If you had not told me, I should have thought that [your remarks on Man] had been added by some one else. As you expected, I differ grievously from you, and I am very sorry for it." Wallace, sensitive to the rebuke, thereafter referred to his theory of human intellect as "my special heresy."

The conventional account of Wallace's apostasy at the brink of complete consistency cites a failure of courage to take the last step and admit man fully into the natural system—a step that Darwin took with commendable fortitude in two books, the *Descent of Man* (1871) and the *Expression of the Emotions* (1872). Thus, Wallace emerges from most historical accounts as a lesser man than Darwin for one (or more) of three reasons, all related to his position on the origins of human intellect: for simple cowardice; for inability to transcend the constraints of culture and traditional views of human uniqueness; and for inconsistency in advocating natural selection so strongly (in the debate on sexual selection), yet abandoning it at the most crucial moment of all.

I cannot analyze Wallace's psyche, and will not comment on his deeper motives for holding fast to the unbridgeable gap between human intellect and the behavior of mere animals. But I can assess the logic of his argument, and recognize that the traditional account of it is not only incorrect, but precisely backwards. Wallace did not abandon natural selection at the human threshold. Rather, it was his peculiarly rigid view of natural selection that led him, quite consistently, to reject it for the human mind. His position never varied—natural selection is the only cause of major evolutionary change. His two debates with Darwin—sexual selection and the origin of human intellect—represent the same argument, not an inconsistent Wallace championing selection in one case and running from it in the other. Wallace's error on human intellect arose from the inadequacy of his rigid selectionism, not from a failure to apply it. And his argument repays our study today, since its flaw persists as the weak link in many of the most "modern" evolutionary speculations of our current literature. For Wallace's rigid selectionism is much closer than Darwin's pluralism to the attitude embodied in our favored theory today, which, ironically in this context, goes by the name of "Neo-Darwinism."

Wallace advanced several arguments for the uniqueness of human intellect, but his central claim begins with an extremely uncommon position for his time, one that commands our highest praise in retrospect. Wallace was one of the few nonracists of the nineteenth century. He really believed that all human groups had innately equal capacities of intellect. Wallace defended his decidedly unconventional egalitarianism with two arguments, anatomical and cultural. He claimed, first of all, that the brains of "savages" are neither much smaller nor more poorly organized than our own: "In the brain of the lowest savages, and, as far as we know, of the prehistoric races, we have an organ . . . little inferior in size and complexity to that of the highest type." Moreover, since cultural conditioning can integrate the rudest savage into our most courtly life, the rudeness itself must arise from a failure to use existing capacities, not from their absence: "It is latent in the lower races, since under European training native military bands have been formed in many parts of the world, which have been able to perform creditably the best modern music."

Of course, in calling Wallace a nonracist, I do not mean to imply that he regarded the cultural practices of all peoples as equal in intrinsic worth. Wallace, like most of his contemporaries, was a cultural chauvinist who did not doubt the evident superiority of European ways. He may have been bullish on the capability of "savages," but he certainly had a low opinion of their life, as he mistook it: "Our law, our government, and our science continually require us to reason through a variety of complicated phenomena to the expected result. Even our games, such as chess, compel us to exercise all these faculties in a remarkable degree. Compare this with the savage languages, which contain no words for abstract conceptions; the utter want of foresight of the savage man beyond his simplest necessities; his inability to combine, or to compare, or to reason on any general subject that does not immediately appeal to his senses."

Hence, Wallace's dilemma: all "savages," from our actual ancestors to modern survivors, had brains fully capable of developing and appreciating all the finest subtleties of European art, morality and philosophy; yet they used, in the state of nature, only the tiniest fraction of that capacity in constructing their rudimentary cultures, with impoverished languages and repugnant morality.

But natural selection can only fashion a feature for immediate use. The brain is vastly overdesigned for what it accomplished in primitive society; thus, natural selection could not have built it:

> A brain one-half larger than that of the gorilla would . . . fully have sufficed for the limited mental development of the savage; and we must therefore admit that the large brain he actually possesses could never have been solely developed by any of those laws of evolution, whose essence is, that they lead to a degree of organization exactly proportionate to the wants of each species, never beyond those wants. . . . Natural selection could only have endowed savage man with a brain a few degrees superior to that of an ape, whereas he actually possesses one very little inferior to that of a philosopher.

Wallace did not confine this general argument to abstract intellect, but extended it to all aspects of European "refinement," to language and music in particular. Consider his views on "the wonderful power, range, flexibility, and sweetness of the musical sounds producible by the human larynx, especially in the female sex."

The habits of savages give no indication of how this faculty could have been developed by natural selection, because it is never required or used by them. The singing of savages is a more or less monotonous howling, and the females seldom sing at all. Savages certainly never choose their wives for fine voices, but for rude health, and strength, and physical beauty. Sexual selection could not therefore have developed this wonderful power, which only comes into play among civilized people. It seems as if the organ had been prepared in anticipation of the future progress in man, since it contains latest capacities which are useless to him in his earlier condition.

Finally, if our higher capacities arose before we used or needed them, then they cannot be the product of natural selection. And, if they originated in anticipation of a future need, then they must be the direct creation of a higher intelligence: "The inference I would draw from this class of phenomena is, that a superior intelligence has guided the development of man in a definite direction, and for a special purpose." Wallace had rejoined the camp of natural theology and Darwin remonstrated, failed to budge his partner, and finally lamented.

The fallacy of Wallace's argument is not a simple unwillingness to extend evolution to humans, but rather the hyper-selectionism that permeated all his evolutionary thought. For if hyper-selectionism is valid—if every part of every creature is fashioned for and only for its immediate use—then Wallace cannot be gainsaid. The earliest Cro-Magnon people, with brains bigger than our own, produced stunning paintings in their caves, but did not write symphonies or build computers. All that we have accomplished since then is the product of cultural evolution based on a brain of unvarying capacity. In Wallace's view, that brain could not be the product of natural selection, since it always possessed capacities so far in excess of its original function.

But hyper-selectionism is not valid. It is a caricature of Darwin's subtler view, and it both ignores and misunderstands the nature of organic form and function. Natural selection may build an organ "for" a specific function or group of functions. But this "purpose" need not fully specify the capacity of that organ. Objects designed for definite purposes can, as a result of their structural complexity, perform many other tasks as well. A factory may install a computer only to issue the monthly pay checks, but such a machine can also analyze the election returns or whip anyone's ass (or at least perpetually tie them) in tic-tack-toe. Our large brains may have originated "for" some set of necessary skills in gathering food, socializing, or whatever; but these skills do not exhaust the limits of what such a complex machine can do. Fortunately for us, those limits include, among other things, an ability to write, from shopping lists for all of us to grand opera for a few. And our larynx may have arisen "for" a limited range of articulated sound needed to coordinate social life. But its physical design permits us to do more with it, from singing in the shower for all to the occasional diva.

Hyper-selectionism has been with us for a long time in various guises; for it represents the late nineteenth century's scientific version of the myth of natural harmony—all is for the best in the best of all possible worlds (all structures well designed for a definite purpose in this case). It is, indeed, the vision of foolish Dr. Pangloss, so vividly satirized by Voltaire in *Candide*—the world is not necessarily good, but it is the best we could possibly have. As the good doctor said in a famous passage that predated Wallace by a century, but captures the essence of what is so deeply wrong with his argument: "Things cannot be other than they are. . . . Everything is made for the best purpose. Our noses were made to carry spectacles, so we have spectacles. Legs were clearly intended for breeches, and we wear them." Nor is Panglossianism dead today—not when so many books in the pop literature on human behavior state that we evolved our big brain "for" hunting and then trace all our current ills to limits of thought and emotion supposedly imposed by such a mode of life.

Ironically then, Wallace's hyper-selectionism led right back to the basic belief of the creationism that it meant to replace—a faith in the "rightness" of things, a definite place for each object in an integrated whole. As Wallace wrote, quite unfairly, of Darwin:

> He whose teachings were at first stigmatized as degrading or even atheistical, by devoting to the varied phenomena of living things the loving, patient, and reverent study of one who really had faith in the beauty and harmony and perfection of creation, was enabled to bring to light innumerable adaptations, and to prove that the most insignificant parts of the meanest living things had a use and a purpose.

I do not deny that nature has its harmonies. But structure also has its latent capacities. Built for one thing, it can do others—and in this flexibility lies both the messiness and the hope of our lives.

5 | Darwin's Middle Road

"WE BEGAN TO sail up the narrow strait lamenting," narrates Odysseus. "For on the one hand lay Scylla, with twelve feet all dangling down; and six necks exceeding long, and on each a hideous head, and therein three rows of teeth set thick and close, full of black death. And on the other mighty Charybdis sucked down the salt sea water. As often as she belched it forth, like a cauldron on a great fire she would seethe up through all her troubled deeps." Odysseus managed to swerve around Charybdis, but Scylla grabbed six of his finest men and devoured them in his sight—"the most pitiful thing mine eyes have seen of all my travail in searching out the paths of the sea."

False lures and dangers often come in pairs in our legends and metaphors—consider the frying pan and the fire, or the devil and the deep blue sea. Prescriptions for avoidance either emphasize a dogged steadiness—the straight and narrow of Christian evangelists—or an averaging between unpleasant alternatives—the golden mean of Aristotle. The idea of steering a course between undesirable extremes emerges as a central prescription for a sensible life.

The nature of scientific creativity is both a perennial topic of discussion and a prime candidate for seeking a golden mean. The two extreme positions have not been directly competing for allegiance of the unwary. They have, rather, replaced each other sequentially, with one now in the ascendency, the other eclipsed.

The first—inductivism—held that great scientists are primarily great observers and patient accumulators of information. For new and significant theory, the inductivists claimed, can only arise from a firm foundation of facts. In this architectural view, each fact is a brick in a structure built without blueprints. Any talk or thought about theory (the completed building) is fatuous and premature before the bricks are set. Inductivism once commanded great prestige within science, and even represented an "official" position of sorts, for it touted, however falsely, the utter honesty, complete objectivity, and almost automatic nature of scientific progress towards final and incontrovertible truth.

Yet, as its critics so rightly claimed, inductivism also depicted science as a heartless, almost inhuman discipline offering no legitimate place to quirkiness, intuition, and all the other subjective attributes adhering to our vernacular notion of genius. Great scientists, the critics claimed, are distinguished more by their powers of hunch and synthesis, than their skill in experiment or observation. The criticisms of inductivism are certainly valid and I welcome its dethroning during the past thirty years as a necessary prelude to better understanding. Yet, in attacking it so strongly, some critics have tried to substitute an alternative equally extreme and unproductive in its emphasis on the essential subjectivity of creative thought. In this "eureka" view, creativity is an ineffable something, accessible only to persons of genius. It arises like a bolt of lightning, unanticipated, unpredictable and unanalyzable—but the bolts strike only a few special people. We ordinary mortals must stand in awe and thanks. (The name refers, of course, to the legendary story of Archimedes running naked through the streets of Syracuse shouting eureka [I have discovered it] when water displaced by his bathing body washed the scales abruptly from his eyes and suggested a method for measuring volumes.)

I am equally disenchanted by both these opposing extremes. Inductivism reduces genius to dull, rote operations; eurekaism grants it an inaccessible status more in the domain of intrinsic mystery than in a realm where we might understand and learn from it. Might we not marry the good features of each view, and abandon both the elitism of eurekaism and the pedestrian qualities of inductivism. May we not acknowledge the personal and subjective character of creativity, but still comprehend it as a mode of thinking that emphasizes or exaggerates capacities sufficiently common to all of us that we may at least understand if not hope to imitate.

In the hagiography of science, a few men hold such high positions that all arguments must apply to them if they are to have any validity. Charles Darwin, as the principal saint of evolutionary biology, has therefore been presented both as an inductivist and as a primary example of eurekaism. I will attempt to show that these interpretations are equally inadequate, and that recent scholarship on Darwin's own odyssey towards the theory of natural selection supports an intermediate position.

So great was the prestige of inductivism in his own day, that Darwin himself fell under its sway and, as an old man, falsely depicted his youthful accomplishments in its light. In an autobiography, written as a lesson in morality for his children and not intended for publication, he penned some famous lines that misled historians for nearly a hundred years. Describing his path to the theory of natural selection, he claimed: "I worked on true Baconian principles, and without any theory collected facts on a wholesale scale."

The inductivist interpretation focuses on Darwin's five years aboard the *Beagle* and explains his transition from a student for the ministry to the nemesis of preachers as the result of his keen powers of observation applied to the whole world. Thus, the traditional story goes, Darwin's eyes opened wider and wider as he saw, in sequence, the bones of giant South American fossil mammals, the turtles and finches of the Galapagos, and the marsupial fauna of Australia. The truth of evolution and its mechanism of natural selection crept up gradually upon him as he sifted facts in a sieve of utter objectivity.

The inadequacies of this tale are best illustrated by the

falsity of its conventional premier example—the so-called Darwin's finches of the Galapagos. We now know that although these birds share a recent and common ancestry on the South American mainland, they have radiated into an impressive array of species on the outlying Galapagos. Few terrestrial species manage to cross the wide oceanic barrier between South America and the Galapagos. But the fortunate migrants often find a sparsely inhabited world devoid of the competitors that limit their opportunities on the crowded mainland. Hence, the finches evolved into roles normally occupied by other birds and developed their famous set of adaptations for feeding—seed crushing, insect eating, even grasping and manipulating a cactus needle to dislodge insects from plants. Isolation—both of the islands from the mainland and among the islands themselves—provided an opportunity for separation, independent adaptation, and speciation.

According to the traditional view, Darwin discovered these finches, correctly inferred their history, and wrote the famous lines in his notebook: "If there is the slightest foundation for these remarks the zoology of Archipelagoes will be worth examining; for such facts would undermine the stability of Species." But, as with so many heroic tales from Washington's cherry tree to the piety of Crusaders, hope rather than truth motivates the common reading. Darwin found the finches to be sure. But he didn't recognize them as variants of a common stock. In fact, he didn't even record the island of discovery for many of them—some of his labels just read "Galapagos Islands." So much for his immediate recognition of the role of isolation in the formation of new species. He reconstructed the evolutionary tale only after his return to London, when a British Museum ornithologist correctly identified all the birds as finches.

The famous quotation from his notebook refers to Galapagos tortoises and to the claim of native inhabitants that they can "at once pronounce from which Island any Tortoise may have been brought" from subtle differences in size and shape of body and scales. This is a statement of different, and much reduced, order from the traditional tale of finches. For the finches are true and separate species—a living example of evolution. The subtle differences among tortoises represent minor geographic variation within a species. It is a jump in reasoning, albeit a valid one as we now know, to argue that such small differences can be amplified to produce a new species. All creationists, after all, acknowledged geographic variation (consider human races), but argued that it could not proceed beyond the rigid limits of a created archetype.

I don't wish to downplay the pivotal influence of the *Beagle* voyage on Darwin's career. It gave him space, freedom and endless time to think in his favored mode of independent self-stimulation. (His ambivalence towards university life, and his middling performance there by conventional standards, reflected his unhappiness with a curriculum of received wisdom.) He writes from South America in 1834: "I have not one clear idea about cleavage, stratification, lines of upheaval. I have no books, which tell me much and what they do I cannot apply to what I see. In consequence I draw my own conclusions, and most gloriously ridiculous ones they are." The rocks and plants and animals that he saw did provoke him to the crucial attitude of doubt—midwife of all creativity. Sydney, Australia—1836. Darwin wonders why a rational God would create so many marsupials on Australia since nothing about its climate or geography suggests any superiority for pouches: "I had been lying on a sunny bank and was reflecting on the strange character of the animals of this country as compared to the rest of the World. An unbeliever in everything beyond his own reason might exclaim, 'Surely two distinct Creators must have been at work.' "

Nonetheless, Darwin returned to London without an evolutionary theory. He suspected the truth of evolution, but had no mechanism to explain it. Natural selection did not arise from any direct reading of the *Beagle*'s facts, but from two subsequent years of thought and struggle as reflected in a series of remarkable notebooks that have been unearthed and published during the past twenty years. In these notebooks, we see Darwin testing and abandoning a

number of theories and pursuing a multitude of false leads
—so much for his later claim about recording facts with an
empty mind. He read philosophers, poets, and economists,
always searching for meaning and insight—so much for the
notion that natural selection arose inductively from the *Beagle*'s facts. Later, he labelled one notebook as "full of metaphysics on morals."

Yet if this tortuous path belies the Scylla of inductivism,
it has engendered an equally simplistic myth—the Charybdis of eurekaism. In his maddeningly misleading autobiography, Darwin does record a eureka and suggests that natural selection struck him as a sudden, serendipitous flash
after more than a year of groping frustration:

> In October 1838, that is, fifteen months after I had
> begun my systematic inquiry, I happened to read for
> amusement Malthus on Population, and being well prepared to appreciate the struggle for existence which
> everywhere goes on from long-continued observation
> of the habits of animals and plants, it at once struck me
> that under these circumstances favorable variations
> would tend to be preserved, and unfavorable ones to
> be destroyed. The result of this would be the formation
> of new species. Here, then, I had at last got a theory by
> which to work.

Yet, again, the notebooks belie Darwin's later recollections—in this case by their utter failure to record, at the
time it happened, any special exultation over his Malthusian
insight. He inscribes it as a fairly short and sober entry
without a single exclamation point, though he habitually
used two or three in moments of excitement. He did not
drop everything and reinterpret a confusing world in its
light. On the very next day, he wrote an even longer passage
on the sexual curiosity of primates.

The theory of natural selection arose neither as a workmanlike induction from nature's facts, nor as a mysterious
bolt from Darwin's subconscious, triggered by an accidental
reading of Malthus. It emerged instead as the result of a

conscious and productive search, proceeding in a ramifying but ordered manner, and utilizing both the facts of natural history and an astonishingly broad range of insights from disparate disciplines far from his own. Darwin trod the middle path between inductivism and eurekaism. His genius is neither pedestrian nor inaccessible.

Darwinian scholarship has exploded since the centennial of the *Origin* in 1959. The publication of Darwin's notebooks and the attention devoted by several scholars to the two crucial years between the *Beagle*'s docking and the demoted Malthusian insight has clinched the argument for a "middle path" theory of Darwin's creativity. Two particularly important works focus on the broadest and narrowest scales. Howard E. Gruber's masterful intellectual and psychological biography of this phase in Darwin's life, *Darwin on Man*, traces all the false leads and turning points in Darwin's search. Gruber shows that Darwin was continually proposing, testing, and abandoning hypotheses, and that he never simply collected facts in a blind way. He began with a fanciful theory involving the idea that new species arise with a prefixed life span, and worked his way gradually, if fitfully, towards an idea of extinction by competition in a world of struggle. He recorded no exultation upon reading Malthus, because the jigsaw puzzle was only missing a piece or two at the time.

Silvan S. Schweber has reconstructed, in detail as minute as the record will allow, Darwin's activities during the few weeks before Malthus (The Origin of the *Origin* Revisited, *Journal of the History of Biology*, 1977). He argues that the final pieces arose not from new facts in natural history, but from Darwin's intellectual wanderings in distant fields. In particular, he read a long review of social scientist and philosopher Auguste Comte's most famous work, the *Cours de philosophie positive*. He was particularly struck by Comte's insistence that a proper theory be predictive and at least potentially quantitative. He then turned to Dugald Stewart's *On the Life and Writing of Adam Smith*, and imbibed the basic belief of the Scottish economists that theories of overall social structure must begin by analyzing the unconstrained actions of individuals. (Natural selection is, above all, a theory about the struggle of individual organisms for success in reproduction.) Then, searching for quantification, he read a lengthy analysis of work by the most famous statistician of his time—the Belgian Adolphe Quetelet. In the review of Quetelet, he found, among other things, a forceful statement of Malthus's quantitative claim—that population would grow geometrically and food supplies only arithmetically, thus guaranteeing an intense struggle for existence. In fact, Darwin had read the Malthusian statement several times before; but only now was he prepared to appreciate its significance. Thus, he did not turn to Malthus by accident, and he already knew what it contained. His "amusement," we must assume, consisted only in a desire to read in its original formulation the familiar statement that had so impressed him in Quetelet's secondary account.

In reading Schweber's detailed account of the moments preceding Darwin's formulation of natural selection, I was particularly struck by the absence of deciding influence from his own field of biology. The immediate precipitators were a social scientist, an economist, and a statistician. If genius has any common denominator, I would propose breadth of interest and the ability to construct fruitful analogies between fields.

In fact, I believe that the theory of natural selection should be viewed as an extended analogy—whether conscious or unconscious on Darwin's part I do not know—to the laissez faire economics of Adam Smith. The essence of Smith's argument is a paradox of sorts: if you want an ordered economy providing maximal benefits to all, then let individuals compete and struggle for their own advantages. The result, after appropriate sorting and elimination of the inefficient, will be a stable and harmonious polity. Apparent order arises naturally from the struggle among individuals, not from predestined principles or higher control. Dugald Stewart epitomized Smith's system in the book Darwin read:

The most effective plan for advancing a people . . . is by allowing every man, as long as he observes the rules of justice, to pursue his own interest in his own way, and to bring both his industry and his capital into the freest competition with those of his fellow citizens. Every system of policy which endeavors . . . to draw towards a particular species of industry a greater share of the capital of the society than would naturally go to it . . . is, in reality, subversive of the great purpose which it means to promote.

As Schweber states: "The Scottish analysis of society contends that the combined effect of individual actions results in the institutions upon which society is based, and that such a society is a stable and evolving one and functions without a designing and directing mind."

We know that Darwin's uniqueness does not reside in his support for the idea of evolution—scores of scientists had preceded him in this. His special contribution rests upon his documentation and upon the novel character of his theory about how evolution operates. Previous evolutionists had proposed unworkable schemes based on internal perfecting tendencies and inherent directions. Darwin advocated a natural and testable theory based on immediate interaction among individuals (his opponents considered it heartlessly mechanistic). The theory of natural selection is a creative transfer to biology of Adam Smith's basic argument for a rational economy: the balance and order of nature does not arise from a higher, external (divine) control, or from the existence of laws operating directly upon the whole, but from struggle among individuals for their own benefits (in modern terms, for the transmission of their genes to future generations through differential success in reproduction).

Many people are distressed to hear such an argument. Does it not compromise the integrity of science if some of its primary conclusions originate by analogy from contemporary politics and culture rather than from data of the discipline itself? In a famous letter to Engels, Karl Marx identified the similarities between natural selection and the English social scene:

It is remarkable how Darwin recognizes among beasts and plants his English society with its division of labor, competition, opening up of new markets, 'invention,' and the Malthusian 'struggle for existence.' It is Hobbes' *bellum omnium contra omnes* (the war of all against all).

Yet Marx was a great admirer of Darwin—and in this apparent paradox lies resolution. For reasons involving all the themes I have emphasized here—that inductivism is inadequate, that creativity demands breadth, and that analogy is a profound source of insight—great thinkers cannot be divorced from their social background. But the source of an idea is one thing; its truth or fruitfulness is another. The psychology and utility of discovery are very different subjects indeed. Darwin may have cribbed the idea of natural selection from economics, but it may still be right. As the German socialist Karl Kautsky wrote in 1902: "The fact that an idea emanates from a particular class, or accords with their interests, of course proves nothing as to its truth or falsity." In this case, it is ironic that Adam Smith's system of laissez faire does not work in his own domain of economics, for it leads to oligopoly and revolution, rather than to order and harmony. Struggle among individuals does, however, seem to be the law of nature.

Many people use such arguments about social context to ascribe great insights primarily to the indefinable phenomenon of good luck. Thus, Darwin was lucky to be born rich, lucky to be on the *Beagle*, lucky to live amidst the ideas of his age, lucky to trip over Parson Malthus—essentially little more than a man in the right place at the right time. Yet, when we read of his personal struggle to understand, the breadth of his concerns and study, and the directedness of his search for a mechanism of evolution, we understand why Pasteur made his famous quip that fortune favors the prepared mind.

6 | Death Before Birth, or a Mite's *Nunc Dimittis*

CAN ANYTHING BE more demoralizing than parental incompetence before the most obvious and innocent of children's questions: why is the sky blue, the grass green? Why does the moon have phases? Our embarrassment is all the more acute because we thought we knew the answer perfectly well, but hadn't rehearsed it since we ourselves had received a bumbled response in similar circumstances a generation earlier. It is the things we think we know—because they are so elementary, or because they surround us—that often present the greatest difficulties when we are actually challenged to explain them.

One such question, with an obvious and incorrect answer, lies close to our biological lives: why, in humans (and in most species familiar to us), are males and females produced in approximately equal numbers? (Actually, males are more common than females at birth in humans, but differential mortality of males leads to a female majority in later life. Still, the departures from a one to one ratio are never great.) At first glance, the answer seems to be, as in Rabelais's motto, "plain as the nose on a man's face." After all, sexual reproduction requires a mate; equal numbers imply universal mating—the happy Darwinian status of maximal reproductive capacity. At second glance, it isn't so clear at all, and we are drawn in confusion to Shakespeare's recasting of the simile: "A jest unseen, inscrutable, invisible, as a nose on a man's face." If maximal reproductive capacity is the optimal state for a species, then why make equal numbers of males and females. Females, after all, set the limit upon numbers of offspring, since eggs are invariably so much larger and less abundant than sperm in species familiar to us—that is, each egg can make an offspring, each sperm cannot. A male can impregnate several females. If a male can mate with nine females and the population contains a hundred individuals, why not make ten males and ninety females? Reproductive capacity will certainly exceed that of a population composed of fifty males and fifty females. Populations made predominantly of females should, by their more rapid rates of reproduction, win any evolutionary race with populations that maintain equality in numbers between the sexes.

What appeared obvious is therefore rendered problematical and the question remains: why do most sexual species contain approximately equal numbers of males and females? The answer, according to most evolutionary biologists, lies in a recognition that Darwin's theory of natural selection speaks only of struggle among *individuals* for reproductive success. It contains no statement about the good of populations, species, or ecosystems. The argument for ninety females and ten males was framed in terms of advantages for populations as a whole—the usual, congenial, and dead wrong, way in which most people think of evolution. If evolution worked for the good of populations as a whole, then sexual species would contain relatively few males.

The observed equality of males and females, in the face of obvious advantages for female predominance if evolution worked upon groups, stands as one of our most elegant demonstrations that Darwin was right—natural selection works by the struggle of individuals to maximize their own reproductive success. The Darwinian argument was first framed by the great British mathematical biologist R.A. Fisher. Suppose, Fisher argued, that either sex began to predominate. Let us say, for example, that fewer males than females are born. Males now begin to leave more offspring than females since their opportunities for mating increase as they become rarer—that is, they impregnate more than one female on average. Thus, if any genetic factors influence the relative proportion of males born to a parent (and such factors do exist), then parents with a genetic inclination to produce males will gain a Darwinian advantage—they will produce more than an average number of grandchildren thanks to the superior reproductive success of their predominantly male offspring. Thus, genes that favor the production of males will spread and male births will rise in frequency. But, this advantage for males fades out as male births increase and it disappears entirely when males equal females in number. Since the same argument works in reverse to favor female births when females are rare, the sex ratio is driven by Darwinian processes to its equilibrium value of one to one.

But how would a biologist go about testing Fisher's theory of sex ratio? Ironically, the species that confirm its predictions are no great help beyond the initial observation. Once we frame the basic argument and determine that the species we know best have approximately equal numbers of males and females, what do we achieve by finding that the next thousand species are similarly ordered? Sure, it all fits, but we do not gain an equal amount of confidence each time we add a new species. Perhaps the one to one ratio exists for another reason?

To test Fisher's theory, we must look for exceptions. We must seek unusual situations in which the premises of Fisher's theory are not met—situations that lead to a specific prediction about how sex ratio should depart from one to one. If change of premises leads to a definite and successful prediction of altered outcome, then we have an independent test that strongly boosts our confidence. This method is embodied in the old proverb that "the exception proves the rule," although many people misunderstand the proverb because it embodies the less common meaning of "prove." Prove comes from the Latin *probare*—to test or to try. Its usual, modern meaning refers to final and convincing demonstration and the motto would seem to say that exceptions establish indubitable validity. But in another sense, closer to its root, "prove" (as in "proving ground" or printer's "proof") is more like its cognate "probe"—a test or an exploration. It is the exception that probes the rule by testing and exploring its consequences in altered situations.

Here nature's rich diversity comes to our aid. The stereotyped image of a birder assiduously adding the rufous-crowned, peg-legged, speckle-backed, cross-billed and cross-eyed towhee to his life list gives, in unwarranted ridicule, a perverted twist to the actual use made by naturalists of life's diversity. It is nature's richness that permits us to establish a science of natural history in the first place—for the variety virtually guarantees that appropriate exceptions can be found to probe any rule. Oddities and weirdnesses are tests of generality, not mere peculiarities to describe and greet with awe or a chuckle.

Fortunately, nature has been profligate in providing species and modes of life that violate the premises of Fisher's argument. In 1967, British biologist W.D. Hamilton (now at the University of Michigan) gathered the cases and arguments into an article entitled "Extraordinary sex ratios." I will discuss in this essay only the clearest and most important of these probing violations.

Nature rarely heeds our homilies in all cases. We are told, and with good reason, that mating of brothers and sisters should be avoided, lest too many unfavorable recessive genes gain an opportunity to express themselves in double dose. (Such genes tend to be rare, and chances are small that two unrelated parents will both carry them. But the probability that two sibs carry the same gene is usually fifty percent.) Nonetheless, some animals never heard the rule and indulge, perhaps exclusively, in sib mating.

Exclusive sib mating destroys the major premise of Fisher's argument for one to one sex ratios. If females are always fertilized by their brothers, then the same parents manufacture both partners of any mating. Fisher assumed that the males had different parents and that an undersupply of males awarded genetic advantages to those parents that could produce males preferentially. But if the same parents produce *both* the mothers and fathers of their grandchildren, then they have an equal genetic investment in each grandchild, no matter what percentage of males and females they produce among their children. In this case, the reason for an equal balance of males and females disappears and the previous argument for female predominance reasserts itself. If each pair of grandparents has a limited store of energy to invest in offspring, and if grandparents producing more offspring gain a Darwinian edge, then grandparents should make as many daughters as possible, and produce only enough sons to ensure that all their daughters will be fertilized. In fact, if their sons can muster sufficient sexual prowess, then parents should make just one son and use every bit of remaining energy to produce as many daughters as they can. As usual, bountiful nature comes to our aid with numerous exceptions to probe Fisher's rule: indeed, species with sib mating also tend to produce a minimal number of males.

Consider the curious life of a male mite in the genus *Adactylidium,* as described by E.A. Albadry and M.S.F. Tawfik in 1966. It emerges from its mother's body and promptly dies within a few hours, having done apparently nothing during its brief life. It attempts, while outside its mother, neither to feed nor to mate. We know about creatures with short adult lives—the mayfly's single day after a much lengthier larval life, for example. But the mayfly mates and insures the continuity of its kind during these few precious hours. The males of *Adactylidium* seem to do nothing at all but emerge and die.

To solve the mystery, we must study the entire life cycle and look inside the mother's body. The impregnated female of *Adactylidium* attaches to the egg of a thrips. That single egg provides the only source of nutrition for rearing all her offspring—for she will feed on nothing else before her death. This mite, so far as we know, engages exclusively in sib mating; thus, it should produce a minimal number of males. Moreover, since total reproductive energy is so strongly constrained by the nutritional resources of a single thrips' egg, progeny are strictly limited, and the more females the better. Indeed, *Adactylidium* matches our prediction by raising a brood of five to eight sisters accompanied by a single male who will serve as both brother and husband to them all. But producing a single male is chancy; if it dies, all sisters will remain virgins and their mother's evolutionary life is over.

If the mite takes a chance on producing but a single male, thus maximizing its potential brood of fertile females, two other adaptations might lessen the risk—providing both protection for the male and guaranteed proximity to his sisters. What better than to rear the brood entirely within a mother's body, feeding both larvae and adults within her, and even allowing copulation to occur inside her protective shell. Indeed, about forty-eight hours after she attaches to the thrips' egg, six to nine eggs hatch within the body of a female *Adactylidium*. The larvae feed on their mother's body, literally devouring her from inside. Two days later, the offspring reach maturity, and the single male copulates with all his sisters. By this time, the mother's tissues have disintegrated, and her body space is a mass of adult mites, their feces, and their discarded larval and nymphal skeletons. The offspring then cut holes through their mother's body wall and emerge. The females must now find a thrips' egg and begin the process again, but the males have already fulfilled their evolutionary role before "birth." They emerge, react however a mite does to the glories of the outside world, and promptly die.

But why not carry the process one stage further? Why should the male be born at all? After copulating with its sisters, its work is done. It is ready to chant the acarine version of Simeon's prayer, *Nunc dimittis*—Oh Lord, now lettest thou thy servant depart in peace. Indeed, since everything that is possible tends to occur at least once in the multifarious world of life, a close relative of *Adactylidium* does just this. *Acarophenax tribolii* also indulges exclusively in sib mating. Fifteen eggs, including but a single male, develop within the mother's body. The male emerges within his mother's shell, copulates with all his sisters and dies before birth. It may not sound like much of a life, but the male *Acarophenax* does as much for its evolutionary continuity as Abraham did in fathering children into his tenth decade.

Nature's oddities are more than good stories. They are material for probing the limits of interesting theories about life's history and meaning.

7 | Shades of Lamarck

THE WORLD, UNFORTUNATELY, rarely matches our hopes and consistently refuses to behave in a reasonable manner. The psalmist did not distinguish himself as an acute observer when he wrote: "I have been young, and now am old; yet have I not seen the righteous forsaken, nor his seed begging bread." The tyranny of what seems reasonable often impedes science. Who before Einstein would have believed that the mass and aging of an object could be affected by its velocity near the speed of light?

Since the living world is a product of evolution, why not suppose that it arose in the simplest and most direct way? Why not argue that organisms improve themselves by their own efforts and pass these advantages to their offspring in the form of altered genes—a process that has long been called, in technical parlance, the "inheritance of acquired characters." This idea appeals to common sense not only for its simplicity but perhaps even more for its happy implication that evolution travels an inherently progressive path, propelled by the hard work of organisms themselves. But, as we all must die, and as we do not inhabit the central body of a restricted universe, so the inheritance of acquired characters represents another human hope scorned by nature.

The inheritance of acquired characters usually goes by the shorter, although historically inaccurate, name of Lamarckism. Jean Baptiste Lamarck (1744–1829), the great French biologist and early evolutionist, believed in the inheritance of acquired characters, but it was not the centerpiece of his evolutionary theory and was certainly not original with him. Entire volumes have been written to trace its pre-Lamarckian pedigree (see Zirkle in bibliography). Lamarck argued that life is generated, continuously and spontaneously, in very simple form. It then climbs a ladder of complexity, motivated by a "force that tends incessantly to complicate organization." This force operates through the creative response of organisms to "felt needs." But life cannot be organized as a ladder because the upward path is often diverted by requirements of local environments; thus, giraffes acquire long necks and wading birds webbed feet, while moles and cave fishes lose their eyes. Inheritance of acquired characters does play an important part in this scheme, but not the central role. It is the mechanism for assuring that offspring benefit from their parents' efforts, but it does not propel evolution up the ladder.

In the late nineteenth century, many evolutionists sought an alternative to Darwin's theory of natural selection. They reread Lamarck, cast aside the guts of it (continuous generation and complicating forces), and elevated one aspect of the mechanics—inheritance of acquired characters—to a central focus it never had for Lamarck himself. Moreover, many of these self-styled "neo-Lamarckians" abandoned Lamarck's cardinal idea that evolution is an active, creative response by organisms to their felt needs. They preserved the inheritance of acquired characters but viewed the acquisitions as direct impositions by impressing environments upon passive organisms.

Although I will bow to contemporary usage and define Lamarckism as the notion that organisms evolve by acquiring adaptive characters and passing them on to offspring in the form of altered genetic information, I do wish to record how poorly this name honors a very fine scientist who died 150 years ago. Subtlety and richness are so often degraded in our world. Consider the poor marshmallow—the plant, that is. Its roots once made a fine candy; now its name adheres to that miserable ersatz of sugar, gelatine, and corn syrup.

Lamarckism, in this sense, remained a popular evolutionary theory well into our century. Darwin won the battle for evolution as a fact, but his theory for its mechanism—natural selection—did not win wide popularity until the traditions of natural history and Mendelian genetics were fused during the 1930s. Moreover, Darwin himself did not deny Lamarckism, although he regarded it as subsidiary to natural selection as an evolutionary mechanism. As late as 1938, for example, Harvard paleontologist Percy Raymond, writing (I suspect) at the very desk I am now using, said of his colleagues: "Probably most are Lamarckians of some shade; to the uncharitable critic it might seem that many out-Lamarck Lamarck." We must recognize the continuing influence of Lamarckism in order to understand much social theory of the recent past—ideas that become incomprehensible if forced into the Darwinian framework we often assume for them. When reformers spoke of the "taint" of poverty, alcoholism, or criminality, they usually thought in quite literal terms—the sins of the father would extend in hard heredity far beyond the third generation. When Lysenko began to advocate Lamarckian cures for the ills of Soviet agriculture during the 1930s, he had not resuscitated a bit of early nineteenth-century nonsense, but a still respectable, if fast fading, theory. Although this tidbit of historical information does not make his hegemony, or the methods he used to retain it, any less appalling, it does render the tale a bit less mysterious. Lysenko's debate with the Russian Mendelians was, at the outset, a legitimate scientific argument. Later, he held on through fraud, deception, manipulation, and murder—that is the tragedy.

Darwin's theory of natural selection is more complex than Lamarckism because it requires *two* separate processes, rather than a single force. Both theories are rooted in the concept of *adaptation*—the idea that organisms respond to changing environments by evolving a form, function, or behavior better suited to these new circumstances. Thus, in both theories, information from the environment must be transmitted to organisms. In Lamarckism, the transfer is direct. An organism perceives the environmental change, responds in the "right" way, and passes its appropriate reaction directly to its offspring.

Darwinism, on the other hand, is a two-step process, with different forces responsible for variation and direction. Darwinians speak of genetic variation, the first step, as "random." This is an unfortunate term because we do not mean random in the mathematical sense of equally likely in all directions. We simply mean that variation occurs with no preferred orientation in adaptive directions. If temperatures are dropping and a hairier coat would aid survival, genetic variation for greater hairiness does not begin to arise with increased frequency. Selection, the second step, works upon *unoriented* variation and changes a population by conferring greater reproductive success upon advantageous variants.

This is the essential difference between Lamarckism and Darwinism—for Lamarckism is, fundamentally, a theory of *directed* variation. If hairy coats are better, animals perceive the need, grow them, and pass the potential to offspring. Thus, variation is directed automatically toward adaptation and no second force like natural selection is needed. Many people do not understand the essential role of directed variation in Lamarckism. They often argue: isn't Lamarckism true because environment does influence heredity—chemical and radioactive mutagens increase the mutation rate and enlarge a population's pool of genetic variation. This mechanism increases the *amount* of variation but does not propel it in favored directions. Lamarckism holds that genetic variation originates *preferentially* in adaptive directions.

In the June 2, 1979, issue of *Lancet,* the leading British medical journal, for example, Dr. Paul E. M. Fine argues for what he calls "Lamarckism" by discussing a variety of biochemical paths for the inheritance of acquired, but *nondirected,* genetic variation. Viruses, essentially naked bits of DNA, may insert themselves into the genetic material of a bacterium and be passed along to offspring as part of the bacterial chromosome. An enzyme called "reverse transcriptase" can mediate the reading of information from cellular RNA "back" into nuclear DNA. The old idea of a single, irreversible flow of information from nuclear DNA through intermediary RNA to proteins that build the body does not hold in all cases—even though Watson himself had once sanctified it as the "central dogma" of molecular biology: DNA makes RNA makes protein. Since an inserted virus is an "acquired character" that can be passed along to offspring, Fine argues that Lamarckism holds in some cases. But Fine has misunderstood the Lamarckian requirement that characters be acquired for adaptive reasons—for Lamarckism is a theory of directed variation. I have heard no evidence that any of these biochemical mechanisms leads to the preferential incorporation of *favorable* genetic information. Perhaps this is possible; perhaps it even happens. If so, it would be an exciting new development, and truly Lamarckian.

But so far, we have found nothing in the workings of Mendelism or in the biochemistry of DNA to encourage a belief that environments or acquired adaptations can direct sex cells to mutate in specific directions. How could colder weather "tell" the chromosomes of a sperm or egg to produce mutations for longer hair? How could Pete Rose transfer hustle to his gametes? It would be nice. It would be simple. It would propel evolution at much faster rates than Darwinian processes allow. But it is not nature's way, so far as we know.

Yet Lamarckism holds on, at least in popular imagination, and we must ask why? Arthur Koestler, in particular, has vigorously defended it in several books, including *The Case of the Midwife Toad,* a full-length attempt to vindicate the Austrian Lamarckian Paul Kammerer, who shot himself in 1926 (although largely for other reasons) after the discovery that his prize specimen had been doctored by an injection of India ink. Koestler hopes to establish at least a "mini-Lamarckianism" to prick the orthodoxy of what he views as a heartless and mechanistic Darwinism. I think that Lamarckism retains its appeal for two major reasons.

First, a few phenomena of evolution do appear, superficially, to suggest Lamarckian explanations. Usually, the Lamarckian appeal arises from a misconception of Darwinism. It is often and truly stated, for example, that many genetic adaptations must be preceded by a shift in behavior without genetic foundation. In a classic and recent case, several species of tits learned to pry the tops off English milk bottles and drink the cream within. One can well imagine a subsequent evolution of bill shape to make the pilferage easier (although it will probably by nipped in the bud by paper cartons and a cessation of home delivery). Is this not Lamarckian in the sense that an active, nongenetic behavioral innovation sets the stage for reinforcing evolution? Doesn't Darwinism think of the environment as a refining fire and organisms as passive entities before it?

But Darwinism is not a mechanistic theory of environmental determinism. It does not view organisms as billiard balls, buffeted about by a shaping environment. These examples of behavioral innovation are thoroughly Darwinian —yet we praise Lamarck for emphasizing so strongly the active role of organisms as creators of their environment. The tits, in learning to invade milk bottles, established new selective pressures by altering their own environment. Bills of a different shape will now be favored by natural selection. The new environment does not provoke the tits to manufacture genetic variation directed toward the favored shape. This, and only this, would be Lamarckian.

Another phenomenon, passing under a variety of names, including the "Baldwin effect" and "genetic assimilation," seems more Lamarckian in character but fits just as well into a Darwinian perspective. To choose the classic illustration: Ostriches have callosities on their legs where they often kneel on hard ground; but the callosities develop within the egg, before they can be used. Does this not require a Lamarckian scenario: Ancestors with smooth legs began to kneel and acquire callosities as a nongenetic adaptation, just as we, depending on our profession, develop writer's calluses or thickened soles. These callosities were then inherited as genetic adaptations, forming well before their use.

The Darwinian explanation for "genetic assimilation"

can be illustrated with the midwife toad of Paul Kammerer, Koestler's favorite example—for Kammerer, ironically, performed a Darwinian experiment without recognizing it. This terrestrial toad descended from aquatic ancestors that grow roughened ridges on their forefeet—the nuptial pads. Males use these pads to hold the female while mating in their slippery environment. Midwife toads, copulating on *terra firma*, have lost the pads, although a few anomalous individuals do develop them in rudimentary form—indicating that the genetic capacity for producing pads has not been entirely lost.

Kammerer forced some midwife toads to breed in water and raised the next generation from the few eggs that had survived in this inhospitable environment. After repeating the process for several generations, Kammerer produced males with nuptial pads (even though one later received an injection of India ink, perhaps not by Kammerer, to heighten the effect). Kammerer concluded that he had demonstrated a Lamarckian effect: he had returned the midwife toad to its ancestral environment; it had reacquired an ancestral adaptation and passed it on in genetic form to offspring.

But Kammerer had really performed a Darwinian experiment: when he forced the toads to breed in water, only a few eggs survived. Kammerer had exerted a strong selection pressure for whatever genetic variation encourages success in water. And he reinforced this pressure over several generations. Kammerer's selection had gathered together the genes that favor aquatic life—a combination that no parent of the first generation possessed. Since nuptial pads are an aquatic adaptation, their expression may be tied to the set of genes that confer success in water—a set enhanced in frequency by Kammerer's Darwinian selection. Likewise, the ostrich may first develop callosities as a nongenetic adaptation. But the habit of kneeling, reinforced by these callosities, also sets up new selective pressures for the preservation of random genetic variation that may also code for these features. The callosities themselves are not mysteriously transferred by inheritance of acquired characters from adult to juvenile.

The second, and I suspect more important reason for Lamarckism's continuing appeal, lies in its offer of some comfort against a universe devoid of intrinsic meaning for our lives. It reinforces two of our deepest prejudices—our belief that effort should be rewarded and our hope for an inherently purposeful and progressive world. Its appeal for Koestler and other humanists lies more with this solace than in any technical argument about heredity. Darwinism offers no such consolation for it holds only that organisms adapt to local environments by struggling to increase their own reproductive success. Darwinism compels us to seek meaning elsewhere—and isn't this what art, music, literature, ethical theory, personal struggle, and Koestlerian humanism are all about? Why make demands of nature and try to restrict her ways when the answers (even if they are personal and not absolute) lie within ourselves?

Thus Lamarckism, so far as we can judge, is false in the domain it has always occupied—as a biological theory of genetic inheritance. Yet, by analogy only, it is the mode of "inheritance" for another and very different kind of "evolution"—human cultural evolution. *Homo sapiens* arose at least 50,000 years ago, and we have not a shred of evidence for any genetic improvement since then. I suspect that the average Cro-Magnon, properly trained, could have handled computers with the best of us (for what it's worth, they had slightly larger brains than we do). All that we have accomplished, for better or for worse, is a result of cultural evolution. And we have done it at rates unmatched by orders of magnitude in all the previous history of life. Geologists cannot measure a few hundred or a few thousand years in the context of our planet's history. Yet, in this millimicrosecond, we have transformed the surface of our planet through the influence of one unaltered biological invention—self-consciousness. From perhaps one hundred thousand people with axes to more than four billion with bombs, rocket ships, cities, televisions, and computers—and all without substantial genetic change.

Cultural evolution has progressed at rates that Darwinian processes cannot begin to approach. Darwinian evolution continues in *Homo sapiens,* but at rates so slow that it no longer has much impact on our history. This crux in the earth's history has been reached because Lamarckian processes have finally been unleashed upon it. Human cultural evolution, in strong opposition to our biological history, is Lamarckian in character. What we learn in one generation, we transmit directly by teaching and writing. Acquired characters are inherited in technology and culture. Lamarckian evolution is rapid and accumulative. It explains the cardinal difference between our past, purely biological mode of change, and our current, maddening acceleration toward something new and liberating—or toward the abyss.

8 | Caring Groups and Selfish Genes

THE WORLD OF objects can be ordered into a hierarchy of ascending levels, box within box. From atoms to molecules made of atoms, to crystals made of molecules, to minerals, rocks, the earth, the solar system, the galaxy made of stars, and the universe of galaxies. Different forces work at different levels. Rocks fall by gravity, but at the atomic and molecular level, gravity is so weak that standard calculations ignore it.

Life, too, operates at many levels, and each has its role in the evolutionary process. Consider three major levels: genes, organisms, and species. Genes are blueprints for organisms; organisms are the building blocks of species. Evolution requires variation, for natural selection cannot operate without a large set of choices. Mutation is the ultimate source of variation, and genes are the unit of variation. Individual organisms are the units of selection. But individuals do not evolve—they can only grow, reproduce, and die. Evolutionary change occurs in groups of interacting organisms; species are the unit of evolution. In short, as philosopher David Hull writes, genes mutate, individuals are selected, and species evolve. Or so the orthodox, Darwinian view proclaims.

The identification of individuals as the unit of selection is a central theme in Darwin's thought. Darwin contended that the exquisite balance of nature had no "higher" cause. Evolution does not recognize the "good of the ecosystem" or even the "good of the species." Any harmony or stability is only an indirect result of individuals relentlessly pursuing their own self-interest—in modern parlance, getting more of their genes into future generations by greater reproductive success. Individuals are the unit of selection; the "struggle for existence" is a matter among individuals.

During the past fifteen years, however, challenges to Darwin's focus on individuals have sparked some lively debates among evolutionists. These challenges have come from above and below. From above, Scottish biologist V.C. Wynne-Edwards raised orthodox hackles fifteen years ago by arguing that groups, not individuals, are units of selection, at least for the evolution of social behavior. From below, English biologist Richard Dawkins has recently raised my hackles with his claim that genes themselves are units of selection, and individuals merely their temporary receptacles.

Wynne-Edwards presented his defense of "group selection" in a long book entitled *Animal Dispersion in Relation to Social Behavior.* He began with a dilemma: Why, if individuals only struggle to maximize their reproductive success, do so many species seem to maintain their populations at a fairly constant level, well matched to the resources available? The traditional Darwinian answer invoked external constraints of food, climate, and predation: only so many can be fed, so the rest starve (or freeze or get eaten), and numbers stabilize. Wynne-Edwards, on the other hand, argued that animals regulate their own populations by gauging the restrictions of their environment and regulating their own reproduction accordingly. He recognized right away that such a theory contravened Darwin's insistence on "individual selection" for it required that many individuals limit or forgo their own reproduction for the good of their group.

Wynne-Edwards postulated that most species are divided into many more-or-less discrete groups. Some groups never evolve a way to regulate their reproduction. Within these groups, individual selection reigns supreme. In good years, populations rise and the groups flourish; in bad years, the groups cannot regulate themselves and face severe crash and even extinction. Other groups develop systems of regulation in which many individuals sacrifice their reproduction for the group's benefit (an impossibility if selection can only favor individuals that seek their own advantage). These groups survive the good and the bad. Evolution is a struggle among groups, not individuals. And groups survive if they regulate their populations by the altruistic acts of individuals. "It is necessary," Wynne-Edwards wrote, "to postulate that social organizations are capable of progressive evolution and perfection as entities in their own right."

Wynne-Edwards reinterpreted most animal behavior in this light. The environment, if you will, prints only so many tickets for reproduction. Animals then compete for tickets through elaborate systems of conventionalized rivalry. In territorial species, each parcel of land contains a ticket and animals (usually males) posture for the parcels. Losers accept gracefully and retreat to peripheral celibacy for the good of all. (Wynne-Edwards, of course, does not impute conscious intent to winners and losers. He imagines that some unconscious hormonal mechanism underlies the good grace of losers.)

In species with dominance hierarchies, tickets are allotted to the appropriate number of places, and animals compete for rank. Competition is by bluff and posture, for animals must not destroy each other by fighting like gladiators. They are, after all, only competing for tickets to benefit the group. The contest is more of a lottery than a test of skills; a distribution of the right number of tickets is far more important than who wins. "The conventionalization of rivalry and the foundation of society are one and the same thing," Wynne-Edwards proclaimed.

But how do animals know the number of tickets? Clearly, they cannot, unless they can census their own populations. In his most striking hypothesis, Wynne-Edwards suggested that flocking, swarming, communal singing, and chorusing evolved through group selection as an effective device for censusing. He included "the singing of birds, the trilling of katydids, crickets and frogs, the underwater sounds of fish, and the flashing of fireflies."

Darwinians came down hard on Wynne-Edwards in the

decade following his book. They pursued two strategies. First, they accepted most of Wynne-Edwards's observations, but reinterpreted them as examples of individual selection. They argued, for example, that *who* wins is what dominance hierarchies and territoriality are all about. If the sex ratio between males and females is near 50:50 and if successful males monopolize several females, then not all males can breed. Everyone competes for the Darwinian prize of passing more genes along. The losers don't walk away with grace, content that their sacrifices increase the common good. They have simply been beaten; with luck, they will win on their next try. The result may be a well-regulated population, but the mechanism is individual struggle.

Virtually all Wynne-Edwards's examples of apparent altruism can be rephrased as tales of individual selfishness. In many flocks of birds, for example, the first individual that spots a predator utters a warning cry. The flock scatters but, according to group selectionists, the crier has saved his flockmates by calling attention to himself—self-destruction (or at least danger) for the good of the flock. Groups with altruist criers prevailed in evolution over all selfish, silent groups, despite the danger to individual altruists. But the debates have brought forth at least a dozen alternatives that interpret crying as beneficial for the crier. The cry may put the flock in random motion, thus befuddling the predator and making it less likely that he will catch anyone, including the crier. Or the crier may wish to retreat to safety but dares not break rank to do it alone, lest the predator detect an individual out of step. So he cries to bring the flock along with him. As the crier, he may be disadvantaged relative to flockmates (or he may not, as the first to safety), but he may still be better off than if he had kept silent and allowed the predator to take someone (perhaps himself) at random.

The second strategy against group selection reinterprets apparent acts of disinterested altruism as selfish devices to propagate genes through surviving kin—the theory of kin selection. Siblings, on average, share half their genes. If you die to save three sibs, you pass on 150 percent of yourself through their reproduction. Again, you have acted for your own evolutionary benefit, if not for your corporeal continuity. Kin selection is a form of Darwinian individual selection.

These alternatives do not disprove group selection, for they merely retell its stories in the more conventional Darwinian mode of individual selection. The dust has yet to settle on this contentious issue but a consensus (perhaps incorrect) seems to be emerging. Most evolutionists would now admit that group selection can occur in certain special situations (species made of many very discrete, socially cohesive groups in direct competition with each other). But they regard such situations as uncommon if only because discrete groups are often kin groups, leading to a preference for kin selection as an explanation for altruism within the group.

Yet, just as individual selection emerged relatively unscarred after its battle with group selection from above, other evolutionists launched an attack from below. Genes, they argue, not individuals are the units of selection. They begin by recasting Butler's famous aphorism that a hen is merely the egg's way of making another egg. An animal, they argue, is only DNA's way of making more DNA. Richard Dawkins has put the case most forcefully in his recent book *The Selfish Gene*. "A body," he writes, "is the genes' way of preserving the genes unaltered."

For Dawkins, evolution is a battle among genes, each seeking to make more copies of itself. Bodies are merely the places where genes aggregate for a time. Bodies are temporary receptacles, survival machines manipulated by genes and tossed away on the geological scrap heap once genes have replicated and slaked their insatiable thirst for more copies of themselves in bodies of the next generation. He writes:

> We are survival machines—robot vehicles blindly programmed to preserve the selfish molecules known as genes. . . .
> They swarm in huge colonies, safe inside gigantic lumbering robots . . . they are in you and me; they

created us, body and mind; and their preservation is the ultimate rationale for our existence.

Dawkins explicitly abandons the Darwinian concept of individuals as units of selection: "I shall argue that the fundamental unit of selection, and therefore of self-interest, is not the species, nor the group, nor even, strictly, the individual. It is the gene, the unit of heredity." Thus, we should not talk about kin selection and apparent altruism. Bodies are not the appropriate units. Genes merely try to recognize copies of themselves wherever they occur. They act only to preserve copies and make more of them. They couldn't care less which body happens to be their temporary home.

I begin my criticism by stating that I am not bothered by what strikes most people as the most outrageous component of these statements—the imputation of conscious action to genes. Dawkins knows as well as you and I do that genes do not plan and scheme; they do not act as witting agents of their own preservation. He is only perpetuating, albeit more colorfully than most, a metaphorical shorthand used (perhaps unwisely) by all popular writers on evolution, including myself (although sparingly, I hope). When he says that genes strive to make more copies of themselves, he means: "selection has operated to favor genes that, by chance, varied in such a way that more copies survived in subsequent generations." The second is quite a mouthful; the first is direct and acceptable as metaphor although literally inaccurate.

Still, I find a fatal flaw in Dawkins's attack from below. No matter how much power Dawkins wishes to assign to genes, there is one thing that he cannot give them—direct visibility to natural selection. Selection simply cannot see genes and pick among them directly. It must use bodies as an intermediary. A gene is a bit of DNA hidden within a cell. Selection views bodies. It favors some bodies because they are stronger, better insulated, earlier in their sexual maturation, fiercer in combat, or more beautiful to behold.

If, in favoring a stronger body, selection acted directly

upon a gene for strength, then Dawkins might be vindicated. If bodies were unambiguous maps of their genes, then battling bits of DNA would display their colors externally and selection might act upon them directly. But bodies are no such thing.

There is no gene "for" such unambiguous bits of morphology as your left kneecap or your fingernail. Bodies cannot be atomized into parts, each constructed by an individual gene. Hundreds of genes contribute to the building of most body parts and their action is channeled through a kaleidoscopic series of environmental influences: embryonic and postnatal, internal and external. Parts are not translated genes, and selection doesn't even work directly on parts. It accepts or rejects entire organisms because suites of parts, interacting in complex ways, confer advantages. The image of individual genes, plotting the course of their own survival, bears little relationship to developmental genetics as we understand it. Dawkins will need another metaphor: genes caucusing, forming alliances, showing deference for a chance to join a pact, gauging probable environments. But when you amalgamate so many genes and tie them together in hierarchical chains of action mediated by environments, we call the resultant object a body.

Moreover, Dawkins's vision requires that genes have an influence upon bodies. Selection cannot see them unless they translate to bits of morphology, physiology, or behavior that make a difference to the success of an organism. Not only do we need a one-to-one mapping between gene and body (criticized in the last paragraph), we also need a one-to-one *adaptive* mapping. Ironically, Dawkins's theory arrived just at a time when more and more evolutionists are rejecting the panselectionist claim that all bits of the body are fashioned in the crucible of natural selection. It may be that many, if not most, genes work equally well (or at least well enough) in all their variants and that selection does not choose among them. If most genes do not present themselves for review, then they cannot be the unit of selection.

I think, in short, that the fascination generated by Dawkins's theory arises from some bad habits of Western scientific thought—from attitudes (pardon the jargon) that we call atomism, reductionism, and determinism. The idea that wholes should be understood by decomposition into "basic" units; that properties of microscopic units can generate and explain the behavior of macroscopic results; that all events and objects have definite, predictable, determined causes. These ideas have been successful in our study of simple objects, made of few components, and uninfluenced by prior history. I'm pretty sure that my stove will light when I turn it on (it did). The gas laws build up from molecules to predictable properties of larger volumes. But organisms are much more than amalgamations of genes. They have a history that matters; their parts interact in complex ways. Organisms are built by genes acting in concert, influenced by environments, translated into parts that selection sees and parts invisible to selection. Molecules that determine the properties of water are poor analogues for genes and bodies. I may not be the master of my fate, but my intuition of wholeness probably reflects a biological truth.

3 | Human Evolution

9 | A Biological Homage to Mickey Mouse

AGE OFTEN turns fire to placidity. Lytton Strachey, in his incisive portrait of Florence Nightingale, writes of her declining years:

Destiny, having waited very patiently, played a queer trick on Miss Nightingale. The benevolence and public spirit of that long life had only been equalled by its acerbity. Her virtue had dwelt in hardness. . . . And now the sarcastic years brought the proud woman her punishment. She was not to die as she had lived. The sting was to be taken out of her; she was to be made soft; she was to be reduced to compliance and complacency.

I was therefore not surprised—although the analogy may strike some people as sacrilegious—to discover that the creature who gave his name as a synonym for insipidity had a gutsier youth. Mickey Mouse turned a respectable fifty last year. To mark the occasion, many theaters replayed his debut performance in *Steamboat Willie* (1928). The original Mickey was a rambunctious, even slightly sadistic fellow. In a remarkable sequence, exploiting the exciting new development of sound, Mickey and Minnie pummel, squeeze, and twist the animals on board to produce a rousing chorus of "Turkey in the Straw." They honk a duck with a tight embrace, crank a goat's tail, tweak a pig's nipples, bang a cow's teeth as a stand-in xylophone, and play bagpipe on her udder.

Christopher Finch, in his semiofficial pictorial history of Disney's work, comments: "The Mickey Mouse who hit the movie houses in the late twenties was not quite the well-behaved character most of us are familiar with today. He was mischievous, to say the least, and even displayed a streak of cruelty." But Mickey soon cleaned up his act, leaving to gossip and speculation only his unresolved relationship with Minnie and the status of Morty and Ferdie. Finch continues: "Mickey . . . had become virtually a national symbol, and as such he was expected to behave properly at all times. If he occasionally stepped out of line, any number of letters would arrive at the Studio from citizens and organizations who felt that the nation's moral well-being was in their hands. . . . Eventually he would be pressured into the role of straight man."

As Mickey's personality softened, his appearance changed. Many Disney fans are aware of this transformation through time, but few (I suspect) have recognized the coor-

Mickey's evolution during 50 years (left to right). As Mickey became increasingly well behaved over the years, his appearance became more youthful. Measurements of three stages in his development revealed a larger relative head size, larger eyes, and an enlarged cranium—all traits of juvenility. © Walt Disney Productions

dinating theme behind all the alterations—in fact, I am not sure that the Disney artists themselves explicitly realized what they were doing, since the changes appeared in such a halting and piecemeal fashion. In short, the blander and inoffensive Mickey became progressively more juvenile in appearance. (Since Mickey's chronological age never altered—like most cartoon characters he stands impervious to the ravages of time—this change in appearance at a constant age is a true evolutionary transformation. Progressive juvenilization as an evolutionary phenomenon is called neoteny. More on this later.)

The characteristic changes of form during human growth have inspired a substantial biological literature. Since the head-end of an embryo differentiates first and grows more rapidly in utero than the foot-end (an antero-posterior gradient, in technical language), a newborn child possesses a relatively large head attached to a medium-sized body with diminutive legs and feet. This gradient is reversed through

growth as legs and feet overtake the front end. Heads continue to grow but so much more slowly than the rest of the body that relative head size decreases.

In addition, a suite of changes pervades the head itself during human growth. The brain grows very slowly after age three, and the bulbous cranium of a young child gives way to the more slanted, lower-browed configuration of adulthood. The eyes scarcely grow at all and relative eye size declines precipitously. But the jaw gets bigger and bigger. Children, compared with adults, have larger heads and eyes, smaller jaws, a more prominent, bulging cranium, and smaller, pudgier legs and feet. Adult heads are altogether more apish, I'm sorry to say.

Mickey, however, has traveled this ontogenetic pathway in reverse during his fifty years among us. He has assumed an ever more childlike appearance as the ratty character of *Steamboat Willie* became the cute and inoffensive host to a magic kingdom. By 1940, the former tweaker of pig's nipples gets a kick in the ass for insubordination (as the *Sorcerer's Apprentice* in *Fantasia*). By 1953, his last cartoon, he has gone fishing and cannot even subdue a squirting clam.

The Disney artists transformed Mickey in clever silence, often using suggestive devices that mimic nature's own changes by different routes. To give him the shorter and pudgier legs of youth, they lowered his pants line and covered his spindly legs with a baggy outfit. (His arms and legs also thickened substantially—and acquired joints for a floppier appearance.) His head grew relatively larger and its features more youthful. The length of Mickey's snout has not altered, but decreasing protrusion is more subtly suggested by a pronounced thickening. Mickey's eye has grown in two modes: first, by a major, discontinuous evolutionary shift as the entire eye of ancestral Mickey became the pupil of his descendants, and second, by gradual increase thereafter.

Mickey's improvement in cranial bulging followed an interesting path since his evolution has always been constrained by the unaltered convention of representing his head as a circle with appended ears and an oblong snout.

The "Evolution" of Mickey Mouse

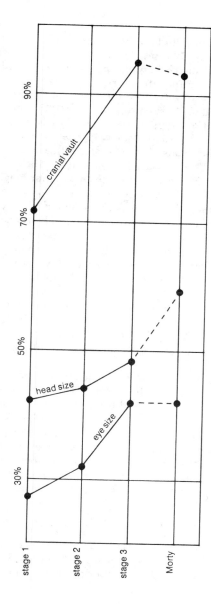

At an early stage in his evolution, Mickey had a smaller head, cranial vault, and eyes. He evolved toward the characteristics of his young nephew Morty (connected to Mickey by a dotted line).

The circle's form could not be altered to provide a bulging cranium directly. Instead, Mickey's ears moved back, increasing the distance between nose and ears, and giving him a rounded, rather than a sloping, forehead.

To give these observations the cachet of quantitative science, I applied my best pair of dial calipers to three stages of the official phylogeny—the thin-nosed, ears-forward figure of the early 1930s (stage 1), the latter-day Jack of Mickey and the Beanstalk (1947, stage 2), and the modern mouse (stage 3). I measured three signs of Mickey's creeping juvenility: increasing eye size (maximum height) as a percentage of head length (base of the nose to top of rear ear); increasing head length as a percentage of body length; and increasing cranial vault size measured by rearward displacement of the front ear (base of the nose to top of front ear as a percentage of base of the nose to top of rear ear).

All three percentages increased steadily—eye size from 27 to 42 percent of head length; head length from 42.7 to 48.1 percent of body length; and nose to front ear from 71.7 to a whopping 95.6 percent of nose to rear ear. For comparison, I measured Mickey's young "nephew" Morty Mouse. In each case, Mickey has clearly been evolving toward youthful stages of his stock, although he still has a way to go for head length.

You may, indeed, now ask what an at least marginally respectable scientist has been doing with a mouse like that. In part, fiddling around and having fun, of course. (I still prefer *Pinocchio* to *Citizen Kane*.) But I do have a serious point—two, in fact—to make. We must first ask why Disney chose to change his most famous character so gradually and persistently in the same direction? National symbols are not altered capriciously and market researchers (for the doll industry in particular) have spent a good deal of time and practical effort learning what features appeal to people as cute and friendly. Biologists also have spent a great deal of time studying a similar subject in a wide range of animals.

In one of his most famous articles, Konrad Lorenz argues that humans use the characteristic differences in form between babies and adults as important behavioral cues. He believes that features of juvenility trigger "innate releasing mechanisms" for affection and nurturing in adult humans. When we see a living creature with babyish features, we feel an automatic surge of disarming tenderness. The adaptive value of this response can scarcely be questioned, for we must nurture our babies. Lorenz, by the way, lists among his releasers the very features of babyhood that Disney affixed progressively to Mickey: "a relatively large head, predominance of the brain capsule, large and low-lying eyes, bulging cheek region, short and thick extremities, a springy elastic consistency, and clumsy movements." (I propose to leave aside for this article the contentious issue of whether or not our affectionate response to babyish features is truly innate and inherited directly from ancestral primates—as Lorenz argues—or whether it is simply learned from our immediate experience with babies and grafted upon an evolutionary predisposition for attaching ties of affection to certain learned signals. My argument works equally well in either case for I only claim that babyish features tend to elicit strong feelings of affection in adult humans, whether the biological basis be direct programming or the capacity to learn and fix upon signals. I also treat as collateral to my point the major thesis of Lorenz's article—that we respond not to the totality or *Gestalt,* but to a set of specific features acting as releasers. This argument is important to Lorenz because he wants to argue for evolutionary identity in modes of behavior between other vertebrates and humans, and we know that many birds, for example, often respond to abstract features rather than *Gestalten.* Lorenz' article, published in 1950, bears the title *Ganzheit und Teil in der tierischen und menschlichen Gemeinschaft*—"Entirety and part in animal and human society." Disney's piecemeal change of Mickey's appearance does make sense in this context—he operated in sequential fashion upon Lorenz's primary releasers.)

Lorenz emphasizes the power that juvenile features hold over us, and the abstract quality of their influence, by pointing out that we judge other animals by the same criteria— although the judgment may be utterly inappropriate in an evolutionary context. We are, in short, fooled by an evolved response to our own babies, and we transfer our reaction to the same set of features in other animals.

Many animals, for reasons having nothing to do with the inspiration of affection in humans, possess some features also shared by human babies but not by human adults—large eyes and a bulging forehead with retreating chin, in particular. We are drawn to them, we cultivate them as pets, we stop and admire them in the wild—while we reject their small-eyed, long-snouted relatives who might make more affectionate companions or objects of admiration. Lorenz points out that the German names of many animals with features mimicking human babies end in the diminutive suffix *chen,* even though the animals are often larger than close relatives without such features—*Rotkehlchen* (robin), *Eichhörnchen* (squirrel), and *Kaninchen* (rabbit), for example.

In a fascinating section, Lorenz then enlarges upon our capacity for biologically inappropriate response to other animals, or even to inanimate objects that mimic human features. "The most amazing objects can acquire remarkable, highly specific emotional values by 'experiential attachment' of human properties. . . . Steeply rising, somewhat overhanging cliff faces or dark storm-clouds piling up have the same, immediate display value as a human being who is standing at full height and leaning slightly forwards" —that is, threatening.

We cannot help regarding a camel as aloof and unfriendly because it mimics, quite unwittingly and for other reasons, the "gesture of haughty rejection" common to so many human cultures. In this gesture, we raise our heads, placing our nose above our eyes. We then half-close our eyes and blow out through our nose—the "harumph" of the stereotyped upperclass Englishman or his well-trained servant. "All this," Lorenz argues quite cogently, "symbolizes resistance against all sensory modalities emanating from the disdained counterpart." But the poor camel cannot help carrying its nose above its elongate eyes, with mouth drawn down. As Lorenz reminds us, if you wish to know whether

Humans feel affection for animals with juvenile features: large eyes, bulging craniums, retreating chins (left column).
Small-eyed, long-snouted animals (right
column) do not elicit the same response.
From *Studies in Animal and Human Behavior*,
vol. II, by Konrad Lorenz, 1971. Methuen
& Co. Ltd.

a camel will eat out of your hand or spit, look at its ears, not the rest of its face.

In his important book *Expression of the Emotions in Man and Animals,* published in 1872, Charles Darwin traced the evolutionary basis of many common gestures to originally adaptive actions in animals later internalized as symbols in humans. Thus, he argued for evolutionary continuity of emotion, not only of form. We snarl and raise our upper lip in fierce anger—to expose our nonexistent fighting canine tooth. Our gesture of disgust repeats the facial actions associated with the highly adaptive act of vomiting in necessary circumstances. Darwin concluded, much to the distress of many Victorian contemporaries: "With mankind some expressions, such as the bristling of the hair under the influence of extreme terror, or the uncovering of the teeth under that of furious rage, can hardly be understood, except on the belief that man once existed in a much lower and animal-like condition."

In any case, the abstract features of human childhood elicit powerful emotional responses in us, even when they occur in other animals. I submit that Mickey Mouse's evolutionary road down the course of his own growth in reverse reflects the unconscious discovery of this biological principle by Disney and his artists. In fact, the emotional status of most Disney characters rests on the same set of distinctions. To this extent, the magic kingdom trades on a biological illusion—our ability to abstract and our propensity to transfer inappropriately to other animals the fitting responses we make to changing form in the growth of our own bodies.

Donald Duck also adopts more juvenile features through time. His elongated beak recedes and his eyes enlarge; he converges on Huey, Louie, and Dewey as surely as Mickey approaches Morty. But Donald, having inherited the mantle of Mickey's original misbehavior, remains more adult in form with his projecting beak and more sloping forehead.

Mouse villains or sharpies, contrasted with Mickey, are always more adult in appearance, although they often share Mickey's chronological age. In 1936, for example, Disney made a short entitled *Mickey's Rival.* Mortimer, a dandy in a yellow sports car, intrudes upon Mickey and Minnie's quiet country picnic. The thoroughly disreputable Mortimer has a head only 29 percent of body length, to Mickey's 45, and a snout 80 percent of head length, compared with Mickey's 49. (Nonetheless, and was it ever different, Minnie transfers her affection until an obliging bull from a neighboring field dispatches Mickey's rival.) Consider also the exaggerated adult features of other Disney characters—the swaggering bully Peg-leg Pete or the simple, if lovable, dolt Goofy.

Dandified, disreputable Mortimer (here stealing Minnie's affections) has strikingly more adult features than Mickey. His head is smaller in proportion to body length; his nose is a full 80 percent of head length. © Walt Disney Productions

As a second, serious biological comment on Mickey's odyssey in form, I note that his path to eternal youth repeats, in epitome, our own evolutionary story. For humans are neotenic. We have evolved by retaining to adulthood the originally juvenile features of our ancestors. Our australopithecine forebears, like Mickey in *Steamboat Willie*, had projecting jaws and low vaulted craniums.

Our embryonic skulls scarcely differ from those of chimpanzees. And we follow the same path of changing form through growth: relative decrease of the cranial vault since brains grow so much more slowly than bodies after birth, and continuous relative increase of the jaw. But while chimps accentuate these changes, producing an adult strikingly different in form from a baby, we proceed much more slowly down the same path and never get nearly so far. Thus, as adults, we retain juvenile features. To be sure, we change enough to produce a notable difference between baby and adult, but our alteration is far smaller than that experienced by chimps and other primates.

A marked slowdown of developmental rates has triggered our neoteny. Primates are slow developers among mammals, but we have accentuated the trend to a degree matched by no other mammal. We have very long periods of gestation, markedly extended childhoods, and the longest life span of any mammal. The morphological features of eternal youth have served us well. Our enlarged brain is, at least in part, a result of extending rapid prenatal growth rates to later ages. (In all mammals, the brain grows rapidly in utero but often very little after birth. We have extended this fetal phase into postnatal life.)

But the changes in timing themselves have been just as important. We are preeminently learning animals, and our extended childhood permits the transference of culture by education. Many animals display flexibility and play in childhood but follow rigidly programmed patterns as adults. Lorenz writes, in the same article cited above: "The characteristic which is so vital for the human peculiarity of the true man—that of always remaining in a state of development—is quite certainly a gift which we owe to the neote-nous nature of mankind."

In short, we, like Mickey, never grow up although we, alas, do grow old. Best wishes to you, Mickey, for your next half-century. May we stay as young as you, but grow a bit wiser.

Cartoon villains are not the only Disney charac-ters with exaggerated adult features. Goofy, like Mortimer, has a small head relative to body length and a prominent snout. © Walt Disney Produc-tions

10 | Piltdown Revisited

NOTHING IS QUITE so fascinating as a well-aged mystery. Many connoisseurs regard Josephine Tey's *The Daughter of Time* as the greatest detective story ever written because its protagonist is Richard III, not the modern and insignificant murderer of Roger Ackroyd. The old chestnuts are perennial sources for impassioned and fruitless debate. Who was Jack the Ripper? Was Shakespeare Shakespeare?

My profession of paleontology offered its entry to the first rank of historical conundrums a quarter-century ago. In 1953, Piltdown man was exposed as a certain fraud perpetrated by a very uncertain hoaxer. Since then, interest has never flagged. People who cannot tell *Tyrannosaurus* from *Allosaurus* have firm opinions about the identity of Piltdown's forger. Rather than simply ask "whodunit?" this column treats what I regard as an intellectually more interesting issue: why did anyone ever accept Piltdown man in the first place? I was led to address the subject by recent and prominent news reports adding—with abysmally poor evidence, in my opinion—yet another prominent suspect to the list. Also, as an old mystery reader, I cannot refrain from expressing my own prejudice, all in due time.

In 1912, Charles Dawson, a lawyer and amateur archeologist from Sussex, brought several cranial fragments to Arthur Smith Woodward, Keeper of Geology at the British Museum (Natural History). The first, he said, had been unearthed by workmen from a gravel pit in 1908. Since then, he had searched the spoil heaps and found a few more fragments. The bones, worn and deeply stained, seemed indigenous to the ancient gravel; they were not the remains of a more recent interment. Yet the skull appeared remarkably modern in form, although the bones were unusually thick.

Smith Woodward, excited as such a measured man could be, accompanied Dawson to Piltdown and there, with Father Teilhard de Chardin, looked for further evidence in the spoil heaps. (Yes, believe it or not, the same Teilhard who, as a mature scientist and theologian, became such a cult figure some fifteen years ago with his attempt to reconcile evolution, nature, and God in *The Phenomenon of Man.* Teilhard had come to England in 1908 to study at the Jesuit College in Hastings, near Piltdown. He met Dawson in a quarry on May 31, 1909; the mature solicitor and the young French Jesuit became warm friends, colleagues, and coexplorers.)

On one of their joint expeditions, Dawson found the famous mandible, or lower jaw. Like the skull fragments, the jaw was deeply stained, but it seemed to be as apish in form as the cranium was human. Nonetheless, it contained two molar teeth, worn flat in a manner commonly encountered in humans, but never in apes. Unfortunately, the jaw was broken in just the two places that might have settled its relationship with the skull: the chin region, with all its marks of distinction between ape and human, and the area of articulation with the cranium.

Armed with skull fragments, the lower jaw, and an associated collection of worked flints and bone, plus a number of mammalian fossils to fix the age as ancient, Smith Woodward and Dawson made their splash before the Geological Society of London on December 18, 1912. Their reception was mixed, although on the whole favorable. No one smelled fraud, but the association of such a human cranium with such an apish jaw indicated to some critics that remains of two separate animals might have been mixed together in the quarry.

During the next three years, Dawson and Smith Woodward countered with a series of further discoveries that, in retrospect, could not have been better programmed to dispel doubt. In 1913, Father Teilhard found the all-important lower canine tooth. It, too, was apish in form but strongly worn in a human manner. Then, in 1915, Dawson convinced most of his detractors by finding the same association of two thick-skulled human cranial fragments with an apish tooth worn in a human manner at a second site two miles from the original finds.

Henry Fairfield Osborn, leading American paleontologist and converted critic, wrote:

> If there is a Providence hanging over the affairs of prehistoric men, it certainly manifested itself in this case, because the three fragments of the second Piltdown man found by Dawson are exactly those which we would have selected to confirm the comparison with the original type. . . . Placed side by side with the corresponding fossils of the first Piltdown man they agree precisely; there is not a shadow of a difference.

Providence, unbeknown to Osborn, walked in human form at Piltdown.

For the next thirty years, Piltdown occupied an uncomfortable but acknowledged place in human prehistory. Then, in 1949, Kenneth P. Oakley applied his fluorine test to the Piltdown remains. Bones pick up fluorine as a function of their time of residence in a deposit and the fluorine content of surrounding rocks and soil. Both the skull and jaw of Piltdown contained barely detectable amounts of fluorine; they could not have lain long in the gravels. Oakley still did not suspect fakery. He proposed that Piltdown, after all, had been a relatively recent interment into ancient gravels.

But a few years later, in collaboration with J.S. Weiner and W.E. le Gros Clark, Oakley finally considered the obvious alternative—that the "interment" had been made in this century with intent to defraud. He found that the skull

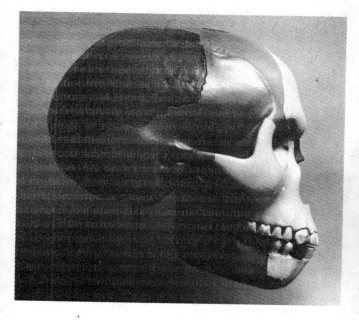

Skull of Piltdown Man.
COURTESY OF THE AMERICAN MUSEUM OF NATURAL HISTORY

and jaw had been artificially stained, the flints and bone worked with modern blades, and the associated mammals, although genuine fossils, imported from elsewhere. Moreover, the teeth had been filed down to simulate human wear. The old anomaly—an apish jaw with a human cranium —was resolved in the most parsimonious way of all. The skull *did* belong to a modern human; the jaw was an orangutan's.

But who had foisted such a monstrous hoax upon scientists so anxious for such a find that they remained blind to an obvious resolution of its anomalies? Of the original trio, Teilhard was dismissed as a young and unwitting dupe. No one has ever (and rightly, in my opinion) suspected Smith Woodward, the superstraight arrow who devoted his life to the reality of Piltdown and who, past eighty and blind, dictated in retirement his last book with its chauvinistic title, *The Earliest Englishman* (1948).

Suspicion instead has focused on Dawson. Opportunity he certainly had, although no one has ever established a satisfactory motive. Dawson was a highly respected amateur with several important finds to his credit. He was overenthusiastic and uncritical, perhaps even a bit unscrupulous in his dealings with other amateurs, but no direct evidence of his complicity has ever come to light. Nevertheless, the circumstantial case is strong and well summarized by J.S. Weiner in *The Piltdown Forgery* (Oxford University Press, 1955).

Supporters of Dawson have maintained that a more professional scientist must have been involved, at least as a coconspirator, because the finds were so cleverly faked. I have always regarded this as a poor argument, advanced by scientists largely to assuage their embarrassment that such an indifferently designed hoax was not detected sooner. The staining, to be sure, had been done consummately. But the "tools" had been poorly carved and the teeth crudely filed—scratch marks were noted as soon as scientists looked with the right hypothesis in mind. Le Gros Clark wrote: "The evidences of artificial abrasion immediately sprang to the eye. Indeed so obvious did they seem it may well be asked—how was it that they had escaped notice before." The forger's main skill consisted in knowing what to leave out—discarding the chin and articulation.

In November 1978, Piltdown reappeared prominently in the news because yet another scientist had been implicated as a possible coconspirator. Shortly before he died at age ninety-three, J.A. Douglas, emeritus professor of geology at Oxford, made a tape recording suggesting that his predecessor in the chair, W.J. Sollas, was the culprit. In support of this assertion, Douglas offered only three items scarcely ranking as evidence in my book: (1) Sollas and Smith Woodward were bitter enemies. (So what. Academia is a den of vipers, but verbal sparring and elaborate hoaxing are responses of differing magnitude.) (2) In 1910, Douglas gave

Sollas some mastodon bones that could have been used as part of the imported fauna. (But such bones and teeth are not rare.) (3) Sollas once received a package of potassium bichromate and neither Douglas nor Sollas's photographer could figure out why he had wanted it. Potassium bichromate was used in staining the Piltdown bones. (It was also an important chemical in photography, and I do not regard the alleged confusion of Sollas's photographer as a strong sign that the professor had some nefarious usages in mind.) In short, I find the evidence against Sollas so weak that I wonder why the leading scientific journals of England and the United States gave it so much space. I would exclude Sollas completely, were it not for the paradox that his famous book, *Ancient Hunters,* supports Smith Woodward's views about Piltdown in terms so obsequiously glowing that it could be read as subtle sarcasm.

Only three hypotheses make much sense to me. First, Dawson was widely suspected and disliked by some amateur archeologists (and equally acclaimed by others). Some compatriots regarded him as a fraud. Others were bitterly jealous of his standing among professionals. Perhaps one of his colleagues devised this complex and peculiar form of revenge. The second hypothesis, and the most probable in my view, holds that Dawson acted alone, whether for fame or to show up the world of professionals we do not know.

The third hypothesis is much more interesting. It would render Piltdown as a joke that went too far, rather than a malicious forgery. It represents the "pet theory" of many prominent vertebrate paleontologists who knew the man well. I have sifted all the evidence, trying hard to knock it down. Instead, I find it consistent and plausible, although not the leading contender. A.S. Romer, late head of the museum I inhabit at Harvard and America's finest vertebrate paleontologist, often stated his suspicions to me. Louis Leakey also believed it. His autobiography refers anonymously to a "second man," but internal evidence clearly implicates a certain individual to anyone in the know.

It is often hard to remember a man in his youth after old age imposes a different persona. Teilhard de Chardin became an austere and almost Godlike figure to many in his later years; he was widely hailed as a leading prophet of our age. But he was once a fun-loving young student. He knew Dawson for three years before Smith Woodward entered the story. He may have had access, from a previous assignment in Egypt, to mammalian bones (probably from Tunisia and Malta) that formed part of the "imported" fauna at Piltdown. I can easily imagine Dawson and Teilhard, over long hours in field and pub, hatching a plot for different reasons: Dawson to expose the gullibility of pompous professionals; Teilhard to rub English noses once again with the taunt that their nation had no legitimate human fossils, while France reveled in a superabundance that made her the queen of anthropology. Perhaps they worked together, never expecting that the leading lights of English science would fasten upon Piltdown with such gusto. Perhaps they expected to come clean but could not.

Teilhard left England to become a stretcher bearer during World War I. Dawson, on this view, persevered and completed the plot with a second Piltdown find in 1915. But then the joke ran away and became a nightmare. Dawson sickened unexpectedly and died in 1916. Teilhard could not return before the war's end. By that time, the three leading lights of British anthropology and paleontology—Arthur Smith Woodward, Grafton Elliot Smith, and Arthur Keith—had staked their careers on the reality of Piltdown. (Indeed they ended up as two Sir Arthurs and one Sir Grafton, largely for their part in putting England on the anthropological map.) Had Teilhard confessed in 1918, his promising career (which later included a major role in describing the legitimate Peking man) would have ended abruptly. So he followed the Psalmist and the motto of Sussex University, later established just a few miles from Piltdown—"Be still, and know. . . ."—to his dying day. Possible. Just possible.

All this speculation provides endless fun and controversy, but what about the prior and more interesting question: why had anyone believed Piltdown in the first place? It was an improbable creature from the start. Why had anyone admitted to our lineage an ancestor with a fully modern cranium and the unmodified jaw of an ape?

Indeed, Piltdown never lacked detractors. Its temporary reign was born in conflict and nurtured throughout by controversy. Many scientists continued to believe that Piltdown was an artifact composed of two animals accidentally commingled in the same deposit. In the early 1940s, for example, Franz Weidenreich, perhaps the world's greatest human anatomist, wrote (with devastating accuracy in hindsight): *"Eoanthropus* ['dawn man,' the official designation of Piltdown] should be erased from the list of human fossils. It is the artificial combination of fragments of a modern human braincase with orang-utanglike mandible and teeth." To this apostasy, Sir Arthur Keith responded with bitter irony: "This is one way of getting rid of facts which do not fit into a preconceived theory; the usual way pursued by men of science is, not to get rid of facts, but frame theory to fit them."

Moreover, had anyone been inclined to pursue the matter, there were published grounds for suspecting fraud from the start. A dental anatomist, C.W. Lyne, stated that the canine found by Teilhard was a young tooth, just erupted before Piltdown's death, and that its intensity of wear could not be reconciled with its age. Others voiced strong doubts about the ancient manufacture of Piltdown's tools. In amateur circles of Sussex, some of Dawson's colleagues concluded that Piltdown must be a fake, but they did not publish their beliefs.

If we are to learn anything about the nature of scientific inquiry from Piltdown—rather than just reveling in the joys of gossip—we will have to resolve the paradox of its easy acceptance. I think that I can identify at least four categories of reasons for the ready welcome accorded to such a misfit by all the greatest English paleontologists. All four contravene the usual mythology about scientific practice—that facts are "hard" and primary and that scientific understanding increases by patient collection and sifting of these objective bits of pure information. Instead, they display science as a human activity, motivated by hope, cultural prejudice, and the pursuit of glory, yet stumbling in its erratic path toward a better understanding of nature.

The imposition of strong hope upon dubious evidence. Before Piltdown, English paleoanthropology was mired in a limbo now occupied by students of extraterrestrial life: endless fields for speculation and no direct evidence. Beyond some flint "cultures" of doubtful human workmanship and some bones strongly suspected as products of recent interments into ancient gravels, England knew nothing of its most ancient ancestors. France, on the other hand, had been blessed with a superabundance of Neanderthals, Cro-Magnons and their associated art and tools. French anthropologists delighted in rubbing English noses with this marked disparity of evidence. Piltdown could not have been better designed to turn the tables. It seemed to predate Neanderthal by a considerable stretch of time. If human fossils had a fully modern cranium hundreds of thousands of years before beetle-browed Neanderthal appeared, then Piltdown must be our ancestor and the French Neanderthals a side branch. Smith Woodward proclaimed: "The Neanderthal race was a degenerate offshoot of early man while surviving modern man may have arisen directly from the primitive source of which the Piltdown skull provides the first discovered evidence." This international rivalry has often been mentioned by Piltdown's commentators, but a variety of equally important factors have usually escaped notice.

Reduction of anomaly by fit with cultural biases. A human cranium with an ape's jaw strikes us today as sufficiently incongruous to merit strong suspicion. Not so in 1913. At that time, many leading paleontologists maintained an a priori preference largely cultural in origin, for "brain primacy" in human evolution. The argument rested on a false inference from contemporary importance to historical priority: we rule today by virtue of our intelligence. Therefore, in our evolution, an enlarged brain must have preceded and inspired all other alterations of our body. We should expect to find human ancestors with enlarged, perhaps nearly modern, brains and a distinctly simian body. (Ironically, nature followed an opposite path. Our earliest ancestors, the australopithecines, were fully erect but still small brained.) Thus, Piltdown neatly matched a widely anticipated result. Grafton Elliot Smith wrote in 1924:

> The outstanding interest of the Piltdown skull is in the confirmation it affords of the view that in the evolution of Man the brain led the way. It is the veriest truism that Man has emerged from the simian state in virtue of the enrichment of the structure of his mind. . . . The brain attained what may be termed the human rank at a time when the jaws and face, and no doubt the body also, still retained much of the uncouthness of Man's simian ancestors. In other words, Man at first . . . was merely an Ape with an overgrown brain. The importance of the Piltdown skull lies in the fact that it affords tangible confirmation of these inferences.

Piltdown also buttressed some all too familiar racial views among white Europeans. In the 1930s and 1940s, following the discovery of Peking man in strata approximately equal in age with the Piltdown gravels, phyletic trees based on Piltdown and affirming the antiquity of white supremacy began to appear in the literature (although they were never adopted by Piltdown's chief champions, Smith Woodward, Smith, and Keith). Peking man (originally called *Sinanthropus*, but now placed in *Homo erectus*) lived in China with a brain two-thirds modern size, while Piltdown man, with its fully developed brain, inhabited England. If Piltdown, as the earliest Englishman, was the progenitor of white races, while other hues must trace their ancestry to *Homo erectus*, then whites crossed the threshold to full humanity long before other people. As longer residents in this exalted state, whites must excel in the arts of civilization.

Reduction of anomaly by matching fact to expectation. We know, in retrospect, that Piltdown had a human cranium and an ape's jaw. As such, it provides an ideal opportunity for testing what scientists do when faced with uncomfortable anomaly. G.E. Smith and others may have advocated an evolutionary head start for the brain, but no one dreamed of an independence so complete that brains might become fully human before jaws changed at all! Piltdown was distressingly too good to be true.

If Keith was right in his taunt to Weidenreich, then Piltdown's champions should have modeled their theories to the uncomfortable fact of a human cranium and an ape's jaw. Instead, they modeled the "facts"—another illustration that information always reaches us through the strong filters of culture, hope, and expectation. As a persistent theme in "pure" description of the Piltdown remains, we learn from all its major supporters that the skull, although remarkably modern, contains a suite of definitely simian characters! Smith Woodward, in fact, originally estimated the cranial capacity at a mere 1,070 cc (compared with a modern average of 1,400 to 1,500), although Keith later convinced him to raise the figure nearer to the low end of our modern spectrum. Grafton Elliot Smith, describing the brain cast in the original paper of 1913, found unmistakable signs of incipient expansion in areas that mark the higher mental faculties in modern brains. He concluded: "We must regard this as being the most primitive and most simian human brain so far recorded; one, moreover, such as might reasonably have been expected to be associated in one and the same individual with the mandible which so definitely indicates the zoological rank of its original possessor." Just a year before Oakley's revelation, Sir Arthur Keith wrote in his last major work (1948): "His forehead was like that of the orang, devoid of a supraorbital torus; in its modeling his frontal bone presented many points of resemblance to that of the orang of Borneo and Sumatra." Modern *Homo sapiens*, I hasten to add, also lacks a supraorbital torus, or brow ridge.

Careful examination of the jaw also revealed a set of remarkably human features for such an apish jaw (beyond the forged wear of the teeth). Sir Arthur Keith repeatedly emphasized, for example, that the teeth were inserted into the jaw in a human, rather than a simian, fashion.

Prevention of discovery by practice. In former years, the British Museum did not occupy the vanguard in maintaining open and accessible collections—a happy trend of recent years, and one that has helped to lift the odor of mustiness (literally and figuratively) from major research museums. Like the stereotype of a librarian who protects books by guarding them from use, Piltdown's keepers severely restricted access to the original bones. Researchers were often permitted to look but not touch; only the set of plaster casts could be handled. Everyone praised the casts for their accuracy of proportion and detail, but the detection of fraud required access to the originals—artificial staining and wear of teeth cannot be discovered in plaster. Louis Leakey writes in his autobiography:

As I write this book in 1972 and ask myself how it was that the forgery remained unmasked for so many years, I have turned my mind back to 1933, when I first went to see Dr. Bather, Smith Woodward's successor. . . . I told him that I wished to make a careful examination of the Piltdown fossils, since I was preparing a textbook on early man. I was taken into the basement to be shown the specimens, which were lifted out of a safe and laid on a table. Next to each fossil was an excellent cast. I was not allowed to handle the originals in any way, but merely to look at them and satisfy myself that the casts were really good replicas. Then, abruptly, the originals were removed and locked up again, and I was left for the rest of the morning with only the casts to study.

It is my belief now that it was under these conditions that all visiting scientists were permitted to examine the Piltdown specimens, and that the situation changed only when they came under the care of my friend and contemporary Kenneth Oakley. He did not see the necessity of treating the fragments as if they were the crown jewels but, rather, considered them simply as important fossils—to be looked after carefully, but from which the maximum scientific evidence should be obtained.

Henry Fairfield Osborn, although not known as a generous man, paid almost obsequious homage to Smith Woodward in his treatise on the historical path of human progress, *Man Rises to Parnassus* (1927). He had been a skeptic
before his visit to the British Museum in 1921. Then, on
Sunday morning, July 24, "after attending a most memora-
ble service in Westminster Abbey," Osborn "repaired to
the British Museum to see the fossil remains of the now
thoroughly vindicated Dawn Man of Great Britain." (He, at
least, as head of the American Museum of Natural History,
got to see the originals.) Osborn swiftly converted and proclaimed Piltdown "a discovery of transcendent importance
to the prehistory of man." He then added: "We have to be
reminded over and over again that Nature is full of para-
doxes and that the order of the universe is not the human
order." Yet Osborn had seen little but the human order on
two levels—the comedy of fraud and the subtler, yet ineluctable, imposition of theory upon nature. Somehow, I am not
distressed that the human order must veil all our interac-
tions with the universe, for the veil is translucent, however
strong its texture.

Postscript

Our fascination with Piltdown never seems to abate. This
article, published originally in March, 1979, elicited a flurry
of correspondence, some acerbic, some congratulatory. It
centered, of course, upon Teilhard. I was not trying to be
cute by writing at length about Teilhard while stating briefly
that Dawson acting alone accounts best for the facts. The
case against Dawson had been made admirably by Weiner,
and I had nothing to add to it. I continued to regard
Weiner's as the most probable hypothesis. But I also believed that the only reasonable alternative (since the second
Piltdown site established Dawson's complicity in my view)
was a coconspiracy—an accomplice for Dawson. The other
current proposals, involving Sollas or even G.E. Smith him-

self, seemed to me so improbable or off-the-wall that I wondered why so little attention had focussed upon the only recognized scientist who had been with Dawson from the start—especially since several of Teilhard's prominent colleagues in vertebrate paleontology harbored private thoughts (or had made cryptically worded public statements) about his possible role.

Ashley Montagu wrote on December 3, 1979, and told me that he had broken the news to Teilhard himself after Oakley's revelation of the fraud—and that Teilhard's astonishment seemed too genuine to represent dissembling: "I feel sure you're wrong about Teilhard. I knew him well, and, in fact, was the first to tell him, the day after it was announced in *The New York Times,* of the hoax. His reaction could hardly have been faked. I have not the slightest doubt that the faker was Dawson." In Paris last September, I spoke with several of Teilhard's contemporaries and scientific colleagues, including Pierre P. Grassé and Jean Piveteau; all regarded any thought of his complicity as monstrous. Père Francois Russo, S.J., later sent me a copy of the letter that Teilhard wrote to Kenneth P. Oakley after Oakley had exposed the fraud. He hoped that this document would assuage my doubts about his coreligionist. Instead my doubts intensified; for, in this letter, Teilhard made a fatal slip. Intrigued by my new role as sleuth, I visited Kenneth Oakley in England on April 16, 1980. He showed me additional documents of Teilhard, and shared other doubts with me. I now believe that the balance of evidence clearly implicatesTeilhard as a coconspirator with Dawson in the Piltdown plot. I will present the entire case in Natural History Magazine in the summer or fall of 1980; but for now, let me mention the internal evidence from Teilhard's first letter to Oakley alone.

Teilhard begins the letter by expressing satisfaction. "I congratulate you most sincerely on your solution of the Piltdown problem . . . I am fundamentally pleased by your conclusions, in spite of the fact that, sentimentally speaking, it spoils one of my brightest and earliest paleontological memories." He continues with his thoughts on "the psychological riddle," or whodunit. he agrees with all others in dismissing Smith Woodward, but he also refuses to implicate Dawson, citing his thorough knowledge of Dawson's character and abilities: "He was a methodical and enthusiastic character . . . In addition, his deep friendship for Sir Arthur makes it almost unthinkable that he should have systematically deceived his associate several years. When we were in the field, I never noticed anything suspicious in his behavior." Teilhard ends by proposing, halfheartedly by his own admission, that the whole affair might have been an accident engendered when an amateur collector threw out some ape bones onto a spoil heap that also contained some human skull fragments, (although Teilhard does not tell us how such a hypothesis could possibly account for the same association two miles away at the second Piltdown site).

Teilhard's slip occurs in his description of the second Piltdown find. Teilhard writes: "He just brought me to the site of Locality 2 and explained me (sic) that he had found the isolated molar and the small pieces of skull in the heaps of rubble and pebbles raked at the surface of the field." Now we know (see Weiner, p. 142) that Dawson did take Teilhard to the second site for a prospecting trip in 1913. He also took Smith Woodward there in 1914. But neither visit led to any discovery; no fossils were found at the second site until 1915. Dawson wrote to Smith Woodward on January 20, 1915 to announce the discovery of two cranial fragments. In July 1915, he wrote again with good news about the discovery of a molar tooth. Smith Woodward assumed (and stated in print) that Dawson had unearthed the specimens in 1915 (see Weiner, p. 144). Dawson became seriously ill later in 1915 and died the next year. Smith Woodward never obtained more precise information from him about the second find. Now, the damning point: Teilhard states explicitly, in the letter quoted above, that Dawson told him about both the tooth and the skull fragments of the second site. But Claude Cuénot, Teilhard's biographer, states that Teilhard was called up for service in December, 1914; and we know that he was at the front on January 22, 1915 (pp. 22–23). But if Dawson did not "officially" discover the molar until July, 1915, how could Teilhard have known about it *unless he was involved in the hoax.* I regard it as unlikely that Dawson would show the material to an innocent Teilhard in 1913 and then withold it from Smith Woodward for two years (especially after taking Smith Woodward to the second site for two days of prospecting in 1914). Teilhard and Smith Woodward were friends and might have compared notes at any time; such an inconsistency on Dawson's part could have blown his cover entirely.

Second, Teilhard states in his letter to Oakley that he did not meet Dawson until 1911: "I knew Dawson very well, since I worked with him and Sir Arthur three or four times at Piltdown (after a chance meeting in a quarry near Hastings in 1911)." Yet it is certain that Teilhard met Dawson during the spring or summer of 1909 (see Weiner, p. 90). Dawson introduced Teilhard to Smith Woodward, and Teilhard submitted some fossils he had found, including a rare tooth of an early mammal, to Smith Woodward late in 1909. When Smith Woodward described this material before the Geological Society of London in 1911, Dawson, in the discussion following Smith Woodward's talk, paid tribute to the "patient and skilled assistance" given to him by Teilhard and another priest since 1909. I don't regard this as a damning point. A first meeting in 1911 would still be early enough for complicity (Dawson "found" his first piece of the Piltdown skull in the autumn of 1911, although he states that a workman had given him a fragment "some years" earlier), and I would never hold a mistake of two years against a man who tried to remember the event forty years later. Still, a later (and incorrect) date, right upon the heels of Dawson's find, certainly averts suspicion.

Moving away from the fascination of whodunit to the theme of my original essay (why did anyone ever believe it in the first place), another colleague sent me an interesting article from *Nature* (the leading scientific periodical in England), November 13, 1913, from the midst of the initial discussions. In it, David Waterston of King's College, University of London, correctly (and definitely) stated that the skull was human, the jaw an ape's. He concludes: "It seems to me to be as inconsequent to refer the mandible and the cranium to the same individual as it would be to articulate a chimpanzee foot with the bones of an essentially human thigh and leg." The correct explanation had been available from the start, but hope, desire, and prejudice prevented its acceptance.

11 | Our Greatest Evolutionary Step

IN MY PREVIOUS book, *Ever Since Darwin,* I began an essay on human evolution with these words:

New and significant prehuman fossils have been unearthed with such unrelenting frequency in recent years that the fate of any lecture notes can only be described with the watchword of a fundamentally irrational economy—planned obsolescence. Each year, when the topic comes up in my courses, I simply open my old folder and dump the contents into the nearest circular file. And here we go again.

And I'm mighty glad I wrote them, because I now want to invoke that passage to recant an argument made later in the same article.

In that essay I reported Mary Leakey's discovery (at Laetoli, thirty miles south of Olduvai Gorge in Tanzania) of the oldest known hominid fossils—teeth and jaws 3.35 to 3.75 million years old. Mary Leakey suggested (and so far as I know, still believes) that these remains should be classified in our genus, *Homo.* I therefore argued that the conventional evolutionary sequence leading from small-brained but fully erect *Australopithecus* to larger-brained *Homo* might have to be reassessed, and that the australopithecines might represent a side branch of the human evolutionary tree.

Early in 1979, newspapers blazed with reports of a new species—more ancient in time and more primitive in appearance than any other hominid fossil—*Australopithecus afarensis*, named by Don Johanson and Tim White. Could any two claims possibly be more different—Mary Leakey's argument that the oldest hominids belong to our own genus, *Homo*, and Johanson and White's decision to name a new species because the oldest hominids possess a set of apelike features shared by no other fossil hominid. Johanson and White must have discovered some new and fundamentally different bones. Not at all. Leakey and Johanson and White are arguing about the same bones. We are witnessing a debate about the interpretation of specimens, not a new discovery.

Johanson worked in the Afar region of Ethiopia from 1972 to 1977 and unearthed an outstanding series of hominid remains. The Afar specimens are 2.9 to 3.3 million years old. Premier among them is the skeleton of an australopithecine named Lucy. She is nearly 40 percent complete—much more than we have ever possessed for any individual from these early days of our history. (Most hominid fossils, even though they serve as a basis for endless speculation and elaborate storytelling, are fragments of jaws and scraps of skulls.)

Johanson and White argue that the Afar specimens and Mary Leakey's Laetoli fossils are identical in form and belong to the same species. They also point out that the Afar and Laetoli bones and teeth represent everything we know about hominids exceeding 2.5 million years in age—all the other African specimens are younger. Finally, they claim that the teeth and skull pieces of these old remains share a set of features absent in later fossils and reminiscent of apes. Thus, they assign the Laetoli and Afar remains to a new species, *A. afarensis*.

The debate is just beginning to warm up, but three opinions have already been vented. Some anthropologists, pointing to different features, regard the Afar and Laetoli specimens as members of our own genus, *Homo*. Others accept Johanson and White's conclusion that these older fossils are closer to the later south and east African *Aus-*

tralopithecus than to *Homo.* But they deny a difference sufficient to warrant a new species and prefer to include the Afar and Laetoli fossils within the species *A. africanus,* originally named for South African specimens in the 1920s. Still others agree with Johanson and White that the Afar and Laetoli fossils deserve a new name.

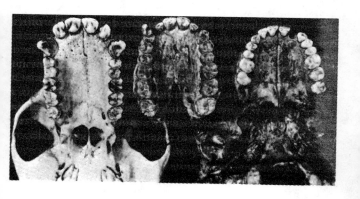

The palate of *Australopithecus afarensis* (center, compared with that of a modern chimpanzee (left) and a human (right). COURTESY OF TIM WHITE AND THE CLEVELAND MUSEUM OF NATURAL HISTORY

As a rank anatomical amateur, my opinion is worth next to nothing. Yet I must say that if a picture is worth all the words of this essay (or only half of them if you follow the traditional equation of 1 for 1,000), the palate of the Afar hominid certainly says "ape" to me. (I must also confess that the designation of *A. afarensis* supports several of my favorite prejudices. Johanson and White emphasize that the Afar and Laetoli specimens span a million years but are virtually identical. I believe that most species do not alter much during the lengthy period of their success and that most evolutionary change accumulates during very rapid events of splitting from ancestral stocks—see essays 17 and 18. Moreover, since I depict human evolution as a bush rather than a ladder, the more species the merrier. Johanson and White do, however, accept far more gradualism than I would advocate for later human evolution.)

Amidst all this argument about skulls, teeth, and taxonomic placement, another and far more interesting feature of the Afar remains has not been disputed. Lucy's pelvis and leg bones clearly show that *A. afarensis* walked as erect as you or I. This fact has been prominently reported by the press, but in a very misleading way. The newspapers have conveyed, almost unanimously, the idea that previous orthodoxy had viewed the evolution of larger brains and upright postures as a gradual transition in tandem, perhaps with brains leading the way—from pea-brained quadrupeds to stooping half brains to fully erect, big-brained *Homo*. The *New York Times* writes (January 1979): "The evolution of bipedalism was thought to have been a gradual process involving intermediate forerunners of modern human beings that were stooped, shuffle-gaited 'ape-men,' creatures more intelligent than apes but not as intelligent as modern human beings." Absolutely false, at least for the past fifty years of our knowledge.

We have known since australopithecines were discovered in the 1920s that these hominids had relatively small brains and fully erect posture. (*A. africanus* has a brain about one-third the volume of ours and a completely upright gait. A correction for its small body size does not remove the large discrepancy between its brain and ours.) This "anomaly" of small brain and upright posture has been a major issue in the literature for decades and wins a prominent place in all important texts.

Thus, the designation of *A. afarensis* does not establish the historical primacy of upright posture over large brains. But it does, in conjunction with two other ideas, suggest something very novel and exciting, something curiously missing from the press reports or buried amidst misinformation about the primacy of upright posture. *A. afarensis* is important because it teaches us that perfected upright gait had already been achieved nearly four million years ago. Lucy's pelvic structure indicates bipedal posture for the Afar remains, while the remarkable footprints just discovered at

Laetoli provide even more direct evidence. The later south and east African australopithecines do not extend back much further than two and a half million years. We have thus added nearly one and a half million years to the history of fully upright posture.

To explain why this addition is so important, I must break the narrative and move to the opposite end of biology—from fossils of whole animals to molecules. During the past fifteen years, students of molecular evolution have accumulated a storehouse of data on the amino acid sequences of similar enzymes and proteins in a wide variety of organisms. This information has generated a surprising result. If we take pairs of species with securely dated times of divergence from a common ancestor in the fossil record, we find that the number of amino acid differences correlates remarkably well with time since the split—the longer that two lineages have been separate, the more the molecular difference. This regularity has led to the establishment of a molecular clock to predict times of divergence for pairs of species without good fossil evidence of ancestry. To be sure, the clock does not beat with the regularity of an expensive watch—it has been called a "sloppy clock" by one of its leading supporters—but it has rarely gone completely haywire.

Darwinians were generally surprised by the clock's regularity because natural selection should work at markedly varying rates in different lineages at different times: very rapidly in complex forms adapting to rapidly changing environments, very slowly in stable, well-adapted populations. If natural selection is the primary cause of evolution in populations, then we should not expect a good correlation between genetic change and time unless rates of selection remain fairly constant—as they should not by the argument stated above. Darwinians have escaped this anomaly by arguing that irregularities in the rate of selection smooth out over long periods of time. Selection might be intense for a few generations and virtually absent for a time thereafter, but the net change averaged over long periods could still be regular. But Darwinians have also been forced to face the possibility that regularity of the molecular clock reflects an evolutionary process not mediated by natural selection, the random fixation of neutral mutations. (I must defer this "hot" topic to another time and more space.)

In any case, the measurement of amino acid differences between humans and living African great apes (gorillas and chimpanzees) led to the most surprising result of all. We are virtually identical for genes that have been studied, despite our pronounced morphological divergence. The average difference in amino acid sequences between humans and African apes is less than one percent (0.8 percent to be precise)—corresponding to a mere five million years since divergence from a common ancestor on the molecular clock. Allowing for the slop, Allan Wilson and Vincent Sarich, the Berkeley scientists who uncovered this anomaly, will accept six million years, but not much more. In short, if the clock is valid, *A. afarensis* is pushing very hard at the theoretical limit of hominid ancestry.

Until recently, anthropologists tended to dismiss the clock, arguing that hominids provided a genuine exception to an admitted rule. They based their skepticism about the molecular clock upon an animal called *Ramapithecus*, an African and Asian fossil known mainly from jaw fragments and ranging back to fourteen million years in age. Many anthropologists claimed that *Ramapithecus* could be placed on our side of the ape-human split—that, in other words, the divergence between hominids and apes occurred more than fourteen million years ago. But this view, based on a series of technical arguments about teeth and their proportions, has been weakening of late. Some of the strongest supporters of *Ramapithecus* as a hominid are now prepared to reassess it as an ape or as a creature near to the common ancestry of ape and human but still before the actual split. The molecular clock has been right too often to cast it aside for some tentative arguments about fragments of jaws. (I now expect to lose a $10 bet I made with Allan Wilson a few years back. He generously gave me seven million years as a maximum for the oldest ape-human common ancestor, but I held out for more. And while I'm not shelling out yet, I don't really expect to collect.*)

We may now put together three points to suggest a major reorientation in views about human evolution: the age and upright posture of *A. afarensis*, the ape-human split on the molecular clock, and the dethroning of *Ramapithecus* as a hominid.

We have never been able to get away from a brain-centered view of human evolution, although it has never represented more than a powerful cultural prejudice imposed upon nature. Early evolutionists argued that enlargement of the brain must have preceded any major alteration of our bodily frame. (See views of G.E. Smith in essay 10. Smith based his pro-Piltdown conviction upon an almost fanatical belief in cerebral primacy.) But *A. africanus*, upright and small brained, ended that conceit in the 1920s, as predicted by a number of astute evolutionists and philosophers, from Ernst Haeckel to Friedrich Engels. Nevertheless, "cerebral primacy," as I like to call it, still held on in altered form. Evolutionists granted the historical primacy of upright posture but conjectured that it arose at a leisurely pace and that the real discontinuity—the leap that made us fully human—occurred much later when, in an unprecedented burst of evolutionary speed, our brains tripled in size within a million years or so.

Consider the following, written ten years ago by a leading expert: "The great leap in cephalization of genus *Homo* took place within the past two million years, after some ten million years of preparatory evolution toward bipedalism, the tool-using hand, etc." Arthur Koestler has carried this view of a cerebral leap toward humanity to an unexcelled height of invalid speculation in his latest book, *Janus*. Our brain grew so fast, he argues, that the outer cerebral cortex, seat of smarts and rationality, lost control over emotive, animal centers deep within our brains. This primitive bestiality surfaces in war, murder, and other forms of mayhem.

---

*Jan., 1980. I just paid. Might as well start off the new decade right.

I believe that we must reassess fundamentally the relative importance we have assigned to upright posture and increase in brain size as determinants of human evolution. We have viewed upright posture as an easily accomplished, gradual trend and increase in brain size as a surprisingly rapid discontinuity—something special both in its evolutionary mode and the magnitude of its effect. I wish to suggest a diametrically opposite view. Upright posture is the surprise, the difficult event, the rapid and fundamental reconstruction of our anatomy. The subsequent enlargement of our brain is, in anatomical terms, a secondary epiphenomenon, an easy transformation embedded in a general pattern of human evolution.

Six million years ago at most, if the molecular clock runs true (and Wilson and Sarich would prefer five), we shared our last common ancestor with gorillas and chimps. Presumably, this creature walked primarily on all fours, although it may have moved about on two legs as well, as apes and many monkeys do today. Little more than a million years later, our ancestors were as bipedal as you or I. This, not later enlargement of the brain, was the great punctuation in human evolution.

Bipedalism is no easy accomplishment. It requires a fundamental reconstruction of our anatomy, particularly of the foot and pelvis. Moreover, it represents an anatomical reconstruction outside the general pattern of human evolution. As I argue in essay 9, through the agency of Mickey Mouse, humans are neotenic—we have evolved by retaining juvenile features of our ancestors. Our large brains, small jaws, and a host of other features, ranging from distribution of bodily hair to ventral pointing of the vaginal canal, are consequences of eternal youth. But upright posture is a different phenomenon. It cannot be achieved by the "easy" route of retaining a feature already present in juvenile stages. For a baby's legs are relatively small and weak, while bipedal posture demands enlargement and strengthening of the legs.

By the time we became upright as *A. afarensis*, the game was largely over, the major alteration of architecture accomplished, the trigger of future change already set. The later enlargement of our brain was anatomically easy. We read our larger brain out of the program of our own growth, by prolonging rapid rates of fetal growth to later times and preserving, as adults, the characteristic proportions of a juvenile primate skull. And we evolved this brain in concert with a host of other neotenic features, all part of a general pattern.

Yet I must end by pulling back and avoiding a fallacy of reasoning—the false equation between magnitude of effect and intensity of cause. As a pure problem in architectural reconstruction, upright posture is far-reaching and fundamental, an enlarged brain superficial and secondary. But the effect of our large brain has far outstripped the relative ease of its construction. Perhaps the most amazing thing of all is a general property of complex systems, our brain prominent among them—their capacity to translate merely quantitative changes in structure into wondrously different qualities of function.

It is now two in the morning and I'm finished. I think I'll walk over to the refrigerator and get a beer; then I'll go to sleep. Culture-bound creature that I am, the dream I will have in an hour or so when I'm supine astounds me ever so much more than the stroll I will now perform perpendicular to the floor.

12 | In the Midst of Life . . .

GREAT STORYTELLERS OFTEN insert bits of humor to relieve the pace of intense drama. Thus, the gravediggers of Hamlet or the courtiers Ping, Pong and Pang of Puccini's *Turandot* prepare us for torture and death to follow. Sometimes, however, episodes that now inspire smiles or laughter were not so designed; the passage of time has obliterated their context and invested the words themselves with an unintended humor in our altered world. Such a passage appears in the midst of geology's most celebrated and serious document—Charles Lyell's *Principles of Geology*, published in three volumes between 1830 and 1833. In it, Lyell argues that the great beasts of yore will return to grace our earth anew:

> Then might those genera of animals return, of which the memorials are preserved in the ancient rocks of our continents. The huge iguanodon might reappear in the woods, and the ichthyosaur in the sea, while the pterodactyl might flit again through the umbrageous groves of tree-ferns.

Lyell's choice of image is striking, but his argument is essential to the major theme of his great work. Lyell wrote the *Principles* to advance his concept of uniformity, his belief that the earth, after "settling down" from the effects of its initial formation, had stayed pretty much the same—no glo-

In a satirical cartoon drawn by one of Lyell's colleagues in response to the cited passage about returning ichthyosaurs and pterodactyls, the future Prof. Ichthyosaurus lectures to students on the skull of a strange creature of the last creation.

bal catastrophes, no steady progress towards any higher state. The extinction of dinosaurs seemed to pose a challenge to Lyell's uniformity. Had they not, after all, been replaced by superior mammals? And didn't this indicate that life's history had a direction? Lyell responded that the replacement of dinosaurs by mammals was part of a grand, recurring cycle—the "great year"—not a step up the ladder of perfection. Climates cycle and life matches climates. Thus, when the summer of the great year came round again, the cold-blooded reptiles would reappear to rule once more.

And yet, for all the fervor of his uniformitarian conviction, Lyell did allow one rather important exception to his vision of an earth marching resolutely in place—the origin of *Homo sapiens* at the latest instant of geological time. Our arrival, he argued, must be viewed as a discontinuity in the history of our planet: "To pretend that such a step, or rather leap, can be part of a regular series of changes in the animal world, is to strain analogy beyond all reasonable bounds." To be sure, Lyell tried to soften the blow he had administered to his own system. He argued that the discontinuity reflected an event in the moral sphere alone—an addition to another realm, not a disruption of the continuing steady-state of the purely material world. The human body, after all, could not be viewed as a Rolls Royce among mammals:

> When it is said that the human race is of far higher dignity than were any pre-existing beings on the earth, it is the intellectual and moral attributes only of our race, not the animal, which are considered; and it is by no means clear, that the organization of man is such as would confer a decided pre-eminence upon him, if, in place of his reasoning powers, he was merely provided with such instincts as are possessed by the lower animals.

Nonetheless, Lyell's argument is a premier example of an all too common tendency among natural historians—the erection of a picket fence around their own species. The fence sports a sign: "so far, but no farther." Again and again, we encounter sweeping visions, encompassing everything from the primordial dust cloud to the chimpanzee. Then, at the very threshold of a comprehensive system, traditional pride and prejudice intervene to secure an exceptional status for one peculiar primate. I discuss another example of the same failure in essay 4—Alfred Russel Wallace's argument for the special creation of human intelligence, the only imposition by divine power upon an organic world constructed entirely by natural selection. The specific form of the argument varies, but its intent is ever the same —to separate man from nature. Below its main sign, Lyell's fence proclaims: "the moral order begins here"; Wallace's reads: "natural selection no longer at work."

Darwin, on the other hand, extended his revolution in thought consistently throughout the entire animal kingdom. Moreover, he explicitly advanced it into the most sensitive areas of human life. Evolution of the human body was upsetting enough, but at least it left the mind potentially inviolate. But Darwin went on. He wrote an entire book to assert that the most refined expressions of human emotion had an animal origin. And if feelings had evolved, could thoughts be far behind?

The picket fence around *Homo sapiens* rests on several supports; the most important posts embody claims for *preparation* and *transcendence*. Humans have not only transcended the ordinary forces of nature, but all that came before was, in some important sense, a preparation for our eventual appearance. Of these two arguments, I regard preparation as by far the more dubious and more expressive of enduring prejudices that we should strive to shed.

Transcendence, in modern guise, states that the history of our peculiar species has been directed by processes that had not operated before on earth. As I argue in essay 7, cultural evolution is our primary innovation. It works by the transmission of skills, knowledge and behavior through learning—a cultural inheritance of acquired characters. This nonbiological process operates in the rapid "Lamarckian" mode, while biological change must plod along by Darwinian steps, glacially slow in comparison. I do not regard this unleashing of Lamarckian processes as a transcendence in the usual sense of overcoming. Biological evolution is neither cancelled nor outmaneuvered. It continues as before and it constrains patterns of culture; but it is too slow to have much impact on the frenetic pace of our changing civilizations.

Preparation, on the other hand, is hubris of a much deeper kind. Transcendence does not compel us to view four billion years of antecedent history as any foreshadowing of our special skills. We may be here by unpredictable good fortune and still embody something new and powerful. But preparation leads us to trace the germ of our later arrival into all previous ages of an immensely long and complicated history. For a species that has been on earth for about 1/100,000 of its existence (fifty thousand of nearly five billion years), this is unwarranted self-inflation of the highest order.

Lyell and Wallace both preached a form of preparation; virtually all builders of picket fences have done so. Lyell depicted an earth in steady-state waiting, indeed almost yearning, for the arrival of a conscious being that could understand and appreciate its sublime and uniform design. Wallace, who turned to spiritualism later in life, advanced the more common claim that physical evolution had occurred in order, ultimately, to link pre-existing mind with a body capable of using it:

> We, who accept the existence of a spiritual world, can look upon the universe as a grand consistent whole adapted in all its parts to the development of spiritual beings capable of indefinite life and perfectibility. To us, the whole purpose, the only *raison d'être* of the world —with all its complexities of physical structure, with its grand geological progress, the slow evolution of the vegetable and animal kingdoms, and the ultimate appearance of man—was the development of the human spirit in association with the human body.

I think that all evolutionists would now reject Wallace's version of the argument for preparation—the foreordination of man in the literal sense. But can there be a legitimate and modern form of the general claim? I believe that such an argument can be constructed, and I also believe that it is the wrong way to view the history of life.

The modern version chucks foreordination in favor of predictability. It abandons the idea that the germ of *Homo sapiens* lay embedded in the primordial bacterium, or that some spiritual force superintended organic evolution, waiting to infuse mind into the first body worthy of receiving it. Instead, it holds that the fully natural process of organic evolution follows certain paths because its primary agent, natural selection, constructs ever more successful designs that prevail in competition against earlier models. The pathways of improvement are rigidly limited by the nature of building materials and the earth's environment. There are only a few ways—perhaps only one—to construct a good flyer, swimmer, or runner. If we could go back to that primordial bacterium and start the process again, evolution would follow roughly the same path. Evolution is more like turning a ratchet than casting water on a broad and uniform slope. It proceeds in a kind of lock step; each stage raises the process one step up, and each is a necessary prelude to the next.

Since life began in microscopic chemistry and has now reached consciousness, the ratchet contains a long sequence of steps. These steps may not be "preparations" in the old sense of foreordination, but they are both predictable and necessary stages in an unsurprising sequence. In an important sense, they prepare the way for human evolution. We are here for a reason after all, even though that reason lies in the mechanics of engineering rather than in the volition of a deity.

But if evolution proceeded as a lock step, then the fossil record should display a pattern of gradual and sequential advance in organization. It does not, and I regard this failure as the most telling argument against an evolutionary ratchet. As I argue in essay 21, life arose soon after the earth itself formed; it then plateaued for as long as three billion years—perhaps five-sixths of its total history. Throughout this vast period, life remained on the prokaryotic level—bacterial and blue green algal cells without the internal structures (nucleus, mitochondria, and others) that make sex and complex metabolism possible. For perhaps three billion years, the highest form of life was an algal mat—thin layers of prokaryotic algae that trap and bind sediment. Then, about 600 million years ago, virtually all the major designs of animal life appeared in the fossil record within a few million years. We do not know why the "Cambrian explosion" occurred when it did, but we have no reason to think that it had to happen then or had to happen at all.

Some scientists have argued that low oxygen levels prevented a previous evolution of complex animal life. If this were true, the ratchet might still work. The stage remained set for three billion years. The screw had to turn in a certain way, but it needed oxygen and had to wait until prokaryotic photosynthesizers gradually supplied the precious gas that the earth's original atmosphere had lacked. Indeed, oxygen was probably rare or absent in the earth's original atmosphere, but it now appears that large amounts had been generated by photosynthesis more than a billion years before the Cambrian explosion.

Thus, we have no reason to regard the Cambrian explosion as more than a fortunate event that need not have occurred, either at all or in the way it did. It may have been a consequence of the evolution of the eukaryotic (nucleate) cell from a symbiotic association of prokaryotic organisms within a single membrane. It may have occurred because the eukaryotic cell could evolve efficient sexual reproduction, and sex distributes and rearranges the genetic variability that Darwinian processes require. But the crucial point is this: if the Cambrian explosion could have occurred any time during more than a billion years before the actual event—that is, for about twice the amount of time that life has spent evolving since then—a ratchet scarcely seems to be an appropriate metaphor for life's history.

If we must deal in metaphors, I prefer a very broad, low and uniform slope. Water drops randomly at the top and usually dries before flowing anywhere. Occasionally, it works its way downslope and carves a valley to channel future flows. These myriad valleys could have arisen anywhere on the landscape. Their current positions are quite accidental. If we could repeat the experiment, we might obtain no valleys at all, or a completely different system. Yet we now stand at the shore line contemplating the fine spacing of valleys and their even contact with the sea. How easy it is to be misled and to assume that no other landscape could possibly have arisen.

I confess that the metaphor of the landscape contains one weak borrowing from its rival, the ratchet. The initial slope does impart a preferred direction to the water dropping on top—even though almost all drops dry before flowing and can flow, when they do, along millions of paths. Doesn't the initial slope imply weak predictability? Perhaps the realm of consciousness occupies such a long stretch of shoreline that some valley would have to reach it eventually.

But here we encounter another constraint, the one that prompted this essay (though I have been, I confess, a long time in getting to it). Almost all drops dry. It took three billion years for any substantial valley to form on the earth's initial slope. It might have taken six billion, or twelve, or twenty for all we know. If the earth were eternal, we might speak of inevitability. But it is not.

Astrophysicist William A. Fowler argues that the sun will exhaust its central hydrogen fuel after ten to twelve billion years of life. It will then explode and transform to a red giant so large that it will extend past the orbit of Jupiter, thus swallowing the earth. It is an arresting thought—the kind that makes you stop and contemplate, or that sends shivers up and down your spine—to recognize that humans have appeared on earth at just about the halfway point of our planet's existence. If the metaphor of the landscape be valid, with all its randomness and unpredictability, then I think we must conclude that the earth need never have evolved its complex life. It took three billion years to go beyond the algal mat. It might as well have taken five times as long, if only the earth had endured. In other words, if we could run the experiment again, the most spectacular event in the history of our solar system, the explosive exhaustion of its parent, might just as well have had an algal mat as its highest, mute witness.

Alfred Russel Wallace also contemplated the eventual destruction of life on earth (though, in his day, physicists argued that the sun would simply burn out and the earth freeze solid). And he could not accept it. He wrote of "the crushing mental burthen imposed upon those who . . . are compelled to suppose that all the slow growths of our race struggling towards a higher life, all the agony of martyrs, all the groans of victims, all the evil and misery and undeserved suffering of the ages, all the struggles for freedom, all the efforts towards justice, all the aspirations for virtue, and the wellbeing of humanity, shall absolutely vanish." Wallace eventually opted for a conventional Christian solution, the eternity of spiritual life: "Beings . . . possessing latent faculties capable of such noble development, are surely destined for a higher and more permanent existence."

I would venture a different argument. The average species of fossil invertebrate lives five to ten million years, as documented in the fossil record. (The oldest may go back, though I doubt the story myself, more than 200 million years.) Vertebrate species tend to live for shorter times. If we are still here to witness the destruction of our planet some five billion years or more hence, then we will have achieved something so unprecedented in the history of life that we should be willing to sing our swan song with joy— *sic transit gloria mundi.* Of course, we might also fly off in those legions of space ships, only to be condensed a bit later into the next big bang. But then, I never have been a keen student of science fiction.

4 | Science and Politics of Human Differences

13 | Wide Hats and Narrow Minds

IN 1861, FROM February to June, the ghost of Baron Georges Cuvier haunted the Anthropological Society of Paris. The great Cuvier, Aristotle of French biology (an immodest designation from which he did not shrink), died in 1832, but the physical vault of his spirit lived on as Paul Broca and Louis Pierre Gratiolet squared off to debate whether or not the size of a brain has anything to do with the intelligence of its bearer.

In the opening round, Gratiolet dared to argue that the best and brightest could not be recognized by their big heads. (Gratiolet, a confirmed monarchist, was no egalitarian. He merely sought other measures to affirm the superiority of white European males.) Broca, founder of the Anthropological Society and the world's greatest craniometrician, or head measurer, replied that "study of the brains of human races would lose most of its interest and utility" if variation in size counted for nothing. Why, he asked, had anthropologists spent so much time measuring heads if the results had no bearing upon what he regarded as the most important question of all—the relative worth of different peoples:

Among the questions heretofore discussed within the Anthropological Society, none is equal in interest and importance to the question before us now. . . . The great importance of craniology has struck anthropolo-

gists with such force that many among us have neglected the other parts of our science in order to devote
ourselves almost exclusively to the study of skulls.
. . . In such data, we hope to find some information
relevant to the intellectual value of the various human
races.

Broca and Gratiolet battled for five months and through
nearly 200 pages of the published bulletin. Tempers flared.
In the heat of battle, one of Broca's lieutenants struck the
lowest blow of all: "I have noticed for a long time that, in
general, those who deny the intellectual importance of the
brain's volume have small heads." In the end, Broca won,
hands down. During the debate, no item of information had
been more valuable to Broca, none more widely discussed
or more vigorously contended, than the brain of Georges
Cuvier.

Cuvier, the greatest anatomist of his time, the man who
revised our understanding of animals by classifying them
according to function—how they work—rather than by rank
in an anthropocentric scale of lower to higher. Cuvier, the
founder of paleontology, the man who first established the
fact of extinction and who stressed the importance of catastrophes in understanding the history both of life and the
earth. Cuvier, the great statesman who, like Talleyrand,
managed to serve all French governments, from revolution
to monarchy, and die in bed. (Actually, Cuvier passed the
most tumultuous years of the revolution as a private tutor
in Normandy, although he feigned revolutionary sympa-
thies in his letters. He arrived in Paris in 1795 and never
left.) F. Bourdier, a recent biographer, describes Cuvier's
corporeal ontogeny, but his words also serve as a good
metaphor for Cuvier's power and influence: "Cuvier was
short and during the Revolution he was very thin; he became stouter during the Empire; and he grew enormously
fat after the Restoration."

Cuvier's contemporaries marveled at his "massive head."
One admirer affirmed that it "gave to his entire person an
undeniable cachet of majesty and to his face an expression

of profound meditation." Thus, when Cuvier died, his colleagues, in the interests of science and curiosity, decided to open the great skull. On Tuesday, May 15, 1832, at seven o'clock in the morning, a group of the greatest doctors and biologists of France gathered to dissect the body of Georges Cuvier. They began with the internal organs and, finding "nothing very remarkable," switched their attention to Cuvier's skull. "Thus," wrote the physician in charge, "we were about to contemplate the instrument of this powerful intelligence." And their expectations were rewarded. The brain of Georges Cuvier weighed 1,830 grams, more than 400 grams above average and 200 grams larger than any nondiseased brain previously weighed. Unconfirmed reports and uncertain inference placed the brains of Oliver Cromwell, Jonathan Swift, and Lord Byron in the same range, but Cuvier had provided the first direct evidence that brilliance and brain size go together.

Broca pushed his advantage and rested a good part of his case on Cuvier's brain. But Gratiolet probed and found a weak spot. In their awe and enthusiasm, Cuvier's doctors had neglected to save either his brain or his skull. Moreover, they reported no measures on the skull at all. The figure of 1,830 g for the brain could not be checked; perhaps it was simply wrong. Gratiolet sought an existing surrogate and had a flash of inspiration: "All brains are not weighed by doctors," he stated, "but all heads are measured by hatters and I have managed to acquire, from this new source, information which, I dare to hope, will not appear to you as devoid of interest." In short, Gratiolet presented something almost bathetic in comparison with the great man's brain: he had found Cuvier's hat! And thus, for two meetings, some of France's greatest minds pondered seriously the meaning of a worn bit of felt.

Cuvier's hat, Gratiolet reported, measured 21.8 cm in length and 18.0 cm in width. He then consulted a certain M. Puriau, "one of the most intelligent and widely known hatters of Paris." Puriau told him that the largest standard size for hats measured 21.5 by 18.5 cm. Although very few men wore a hat so big, Cuvier was not off scale. Moreover, Gratiolet reported with evident pleasure, the hat was extremely flexible and "softened by very long usage." It had probably not been so large when Cuvier bought it. Moreover, Cuvier had an exceptionally thick head of hair, and he wore it bushy. "This seems to prove quite clearly," Gratiolet proclaimed, "that if Cuvier's head was very large, its size was not absolutely exceptional or unique."

Gratiolet's opponents preferred to believe the doctors and refused to grant much weight to a bit of cloth. More than twenty years later, in 1883, G. Hervé again took up the subject of Cuvier's brain and discovered a missing item: Cuvier's head had been measured after all, but the figures had been omitted from the autopsy report. The skull was big indeed. Shaved of that famous mat of hair, as it was for the autopsy, its greatest circumference could be equaled by only 6 percent of "scientists and men of letters" (measured in life with their hair at that) and zero percent of domestic servants. As for the infamous hat, Hervé pleaded ignorance, but he did cite the following anecdote: "Cuvier had a habit of leaving his hat on a table in his waiting room. It often happened that a professor or a statesman tried it on. The hat descended below their eyes."

Yet, just as the doctrine of more-is-better stood on the verge of triumph, Hervé snatched potential defeat from the jaws of Broca's victory. Too much of a good thing can be as troubling as a deficiency, and Hervé began to worry. Why did Cuvier's brain exceed those of other "men of genius" by so much? He reviewed both details of the autopsy and records of Cuvier's frail early health and constructed a circumstantial case for "transient juvenile hydrocephaly," or water on the brain. If Cuvier's skull had been artificially enlarged by the pressure of fluids early during its growth, then a brain of normal size might simply have expanded—by decreasing in density, not by growing larger—into the space available. Or did an enlarged space permit the brain to grow to an unusual size after all? Hervé could not resolve this cardinal question because Cuvier's brain had been measured and then tossed out. All that remained was the magisterial number, 1,830 grams. "With the brain of

Cuvier," wrote Hervé, "science has lost one of the most precious documents it ever possessed."

On the surface, this tale seems ludicrous. The thought of France's finest anthropologists arguing passionately about the meaning of a dead colleague's hat could easily provoke the most misleading and dangerous inference of all about history—a view of the past as a domain of naïve half-wits, the path of history as a tale of progress, and the present as sophisticated and enlightened.

But if we laugh with derision, we will never understand. Human intellectual capacity has not altered for thousands of years so far as we can tell. If intelligent people invested intense energy in issues that now seem foolish to us, then the failure lies in our understanding of their world, not in their distorted perceptions. Even the standard example of ancient nonsense—the debate about angels on pinheads— makes sense once you realize that theologians were not discussing whether five or eighteen would fit, but whether a pin could house a finite or an infinite number. In certain theological systems, the corporeality or noncorporeality of angels is an important matter indeed.

In this case, a clue to the vital importance of Cuvier's brain for nineteenth-century anthropology lies in the last line of Broca's statement, quoted above: "In such data, we hope to find some information relevant to the intellectual value of the various human races." Broca and his school wanted to show that brain size, through its link with intelligence, could resolve what they regarded as the primary question for a "science of man"—explaining why some individuals and groups are more successful than others. To do this, they separated people according to a priori convictions about their worth—men versus women, whites versus blacks, "men of genius" versus ordinary folks—and tried to demonstrate differences in brain size. The brains of eminent men (literally males) formed an essential link in their argument—and Cuvier was the *crème de la crème*. Broca concluded:

In general, the brain is larger in men than in women, in eminent men than in men of mediocre talent, in superior races than in inferior races. Other things equal, there is a remarkable relationship between the development of intelligence and the volume of the brain.

Broca died in 1880, but disciples continued his catalog of eminent brains (indeed, they added Broca's own to the list —although it weighed in at an undistinguished 1,484 grams). The dissection of famous colleagues became something of a cottage industry among anatomists and anthropologists. E.A. Spitzka, the most prominent American practitioner of the trade, cajoled his eminent friends: "To me the thought of an autopsy is certainly less repugnant than I imagine the process of cadaveric decomposition in the grave to be." The two premier American ethnologists, John Wesley Powell and W J McGee made a wager over who had the larger brain—and Spitzka contracted to resolve the issue for them posthumously. (It was a toss-up. The brains of Powell and McGee differed very little, no more than varying body size might require.)

By 1907, Spitzka could present a tabulation of 115 eminent men. As the list grew in length, ambiguity of results increased apace. At the upper end, Cuvier was finally overtaken when Turgenev broke the 2,000-gram barrier in 1883. But embarrassment and insult stalked the other end. Walt Whitman managed to hear the varied carols of America singing with only 1,282 g. Franz Josef Gall, a founder of phrenology—the original "science" of judging mental worth by the size of localized brain areas—could muster only 1,198 g. Later, in 1924, Anatole France almost halved Turgenev's 2,012 and weighed in at a mere 1,017 g.

Spitzka, nonetheless, was undaunted. In an outrageous example of data selected to conform with a priori prejudice, he arranged, in order, a large brain from an eminent white male, a bushwoman from Africa, and a gorilla. (He could easily have reversed the first two by choosing a larger black and a smaller white.) Spitzka concluded, again invoking the shade of Georges Cuvier: "The jump from a Cuvier or a Thackeray to a Zulu or a Bushman is no greater than from the latter to the gorilla or the orang."

Such overt racism is no longer common among scientists, and I trust that no one would now try to rank races or sexes by the average size of their brains. Yet our fascination with the physical basis of intelligence persists (as it should), and the naïve hope remains in some quarters that size or some other unambiguous external feature might capture the subtlety within. Indeed, the crassest form of more-is-better—using an easily measured quantity to assess improperly a far more subtle and elusive quality—is still with us. And the method that some men use to judge the worth of their penises or their automobiles is still being applied to brains. This essay was inspired by recent reports on the whereabouts of Einstein's brain. Yes, Einstein's brain was removed for study, but a quarter century after his death, the results have not been published. The remaining pieces—others were farmed out to various specialists—now rest in a Mason jar packed in a cardboard box marked "Costa Cider" and housed in an office in Wichita, Kansas. Nothing has been published because nothing unusual has been found. "So far it's fallen within normal limits for a man his age," remarked the owner of the Mason jar.

Did I just hear Cuvier and Anatole France laughing in concert from on high? Are they repeating a famous motto of their native land: *plus ça change, plus c'est la même chose* ("the more things change, the more they remain the same"). The physical structure of the brain must record intelligence in some way, but gross size and external shape are not likely to capture anything of value. I am, somehow, less interested in the weight and convolutions of Einstein's brain than in the near certainty that people of equal talent have lived and died in cotton fields and sweatshops.

14 | Women's Brains

IN THE PRELUDE to *Middlemarch*, George Eliot lamented the unfulfilled lives of talented women:

> Some have felt that these blundering lives are due to the inconvenient indefiniteness with which the Supreme Power has fashioned the natures of women: if there were one level of feminine incompetence as strict as the ability to count three and no more, the social lot of women might be treated with scientific certitude.

Eliot goes on to discount the idea of innate limitation, but while she wrote in 1872, the leaders of European anthropometry were trying to measure "with scientific certitude" the inferiority of women. Anthropometry, or measurement of the human body, is not so fashionable a field these days, but it dominated the human sciences for much of the nineteenth century and remained popular until intelligence testing replaced skull measurement as a favored device for making invidious comparisons among races, classes, and sexes. Craniometry, or measurement of the skull, commanded the most attention and respect. Its unquestioned leader, Paul Broca (1824–80), professor of clinical surgery at the Faculty of Medicine in Paris, gathered a school of disciples and imitators around himself. Their work, so meticulous and apparently irrefutable, exerted great influence and won high esteem as a jewel of nineteenth-century science.

Broca's work seemed particularly invulnerable to refutation. Had he not measured with the most scrupulous care and accuracy? (Indeed, he had. I have the greatest respect for Broca's meticulous procedure. His numbers are sound. But science is an inferential exercise, not a catalog of facts. Numbers, by themselves, specify nothing. All depends upon what you do with them.) Broca depicted himself as an apostle of objectivity, a man who bowed before facts and cast aside superstition and sentimentality. He declared that "there is no faith, however respectable, no interest, however legitimate, which must not accommodate itself to the progress of human knowledge and bend before truth." Women, like it or not, had smaller brains than men and, therefore, could not equal them in intelligence. This fact, Broca argued, may reinforce a common prejudice in male society, but it is also a scientific truth. L. Manouvrier, a black sheep in Broca's fold, rejected the inferiority of women and wrote with feeling about the burden imposed upon them by Broca's numbers:

Women displayed their talents and their diplomas. They also invoked philosophical authorities. But they were opposed by *numbers* unknown to Condorcet or to John Stuart Mill. These numbers fell upon poor women like a sledge hammer, and they were accompanied by commentaries and sarcasms more ferocious than the most misogynist imprecations of certain church fathers. The theologians had asked if women had a soul. Several centuries later, some scientists were ready to refuse them a human intelligence.

Broca's argument rested upon two sets of data: the larger brains of men in modern societies, and a supposed increase in male superiority through time. His most extensive data came from autopsies performed personally in four Parisian hospitals. For 292 male brains, he calculated an average weight of 1,325 grams; 140 female brains averaged 1,144 grams for a difference of 181 grams, or 14 percent of the male weight. Broca understood, of course, that part of this difference could be attributed to the greater height of males. Yet he made no attempt to measure the effect of size alone and actually stated that it cannot account for the entire difference because we know, a priori, that women are not as intelligent as men (a premise that the data were supposed to test, not rest upon):

> We might ask if the small size of the female brain depends exclusively upon the small size of her body. Tiedemann has proposed this explanation. But we must not forget that women are, on the average, a little less intelligent than men, a difference which we should not exaggerate but which is, nonetheless, real. We are therefore permitted to suppose that the relatively small size of the female brain depends in part upon her physical inferiority and in part upon her intellectual inferiority.

In 1873, the year after Eliot published *Middlemarch,* Broca measured the cranial capacities of prehistoric skulls from L'Homme Mort cave. Here he found a difference of only 99.5 cubic centimeters between males and females, while modern populations range from 129.5 to 220.7. Topinard, Broca's chief disciple, explained the increasing discrepancy through time as a result of differing evolutionary pressures upon dominant men and passive women:

> The man who fights for two or more in the struggle for existence, who has all the responsibility and the cares of tomorrow, who is constantly active in combating the environment and human rivals, needs more brain than the woman whom he must protect and nourish, the sedentary woman, lacking any interior occupations, whose role is to raise children, love, and be passive.

In 1879, Gustave Le Bon, chief misogynist of Broca's school, used these data to publish what must be the most vicious attack upon women in modern scientific literature (no one can top Aristotle). I do not claim his views were representative of Broca's school, but they were published in France's most respected anthropological journal. Le Bon concluded:

In the most intelligent races, as among the Parisians, there are a large number of women whose brains are closer in size to those of gorillas than to the most developed male brains. This inferiority is so obvious that no one can contest it for a moment; only its degree is worth discussion. All psychologists who have studied the intelligence of women, as well as poets and novelists, recognize today that they represent the most inferior forms of human evolution and that they are closer to children and savages than to an adult, civilized man. They excel in fickleness, inconstancy, absence of thought and logic, and incapacity to reason. Without doubt there exist some distinguished women, very superior to the average man, but they are as exceptional as the birth of any monstrosity, as, for example, of a gorilla with two heads; consequently, we may neglect them entirely.

Nor did Le Bon shrink from the social implications of his views. He was horrified by the proposal of some American reformers to grant women higher education on the same basis as men:

A desire to give them the same education, and, as a consequence, to propose the same goals for them, is a dangerous chimera. . . . The day when, misunderstanding the inferior occupations which nature has given her, women leave the home and take part in our battles; on this day a social revolution will begin, and everything that maintains the sacred ties of the family will disappear.

Sound familiar?*

I have reexamined Broca's data, the basis for all this derivative pronouncement, and I find his numbers sound but his interpretation ill-founded, to say the least. The data supporting his claim for increased difference through time can be easily dismissed. Broca based his contention on the samples from L'Homme Mort alone—only seven male and six female skulls in all. Never have so little data yielded such far ranging conclusions.

In 1888, Topinard published Broca's more extensive data on the Parisian hospitals. Since Broca recorded height and age as well as brain size, we may use modern statistics to remove their effect. Brain weight decreases with age, and Broca's women were, on average, considerably older than his men. Brain weight increases with height, and his average man was almost half a foot taller than his average woman. I used multiple regression, a technique that allowed me to assess simultaneously the influence of height and age upon brain size. In an analysis of the data for women, I found that, at average male height and age, a woman's brain would weigh 1,212 grams. Correction for height and age reduces Broca's measured difference of 181 grams by more than a third, to 113 grams.

I don't know what to make of this remaining difference because I cannot assess other factors known to influence brain size in a major way. Cause of death has an important effect: degenerative disease often entails a substantial diminution of brain size. (This effect is separate from the decrease attributed to age alone.) Eugene Schreider, also working with Broca's data, found that men killed in accidents had brains weighing, on average, 60 grams more than men dying of infectious diseases. The best modern data I can find (from American hospitals) records a full 100-gram

*When I wrote this essay, I assumed that Le Bon was a marginal, if colorful, figure. I have since learned that he was a leading scientist, one of the founders of social psychology, and best known for a seminal study on crowd behavior, still cited today (La psychologie des foules, 1895), and for his work on unconscious motivation.

difference between death by degenerative arteriosclerosis and by violence or accident. Since so many of Broca's subjects were very elderly women, we may assume that lengthy degenerative disease was more common among them than among the men.

More importantly, modern students of brain size still have not agreed on a proper measure for eliminating the powerful effect of body size. Height is partly adequate, but men and women of the same height do not share the same body build. Weight is even worse than height, because most of its variation reflects nutrition rather than intrinsic size— fat versus skinny exerts little influence upon the brain. Manouvrier took up this subject in the 1880s and argued that muscular mass and force should be used. He tried to measure this elusive property in various ways and found a marked difference in favor of men, even in men and women of the same height. When he corrected for what he called "sexual mass," women actually came out slightly ahead in brain size.

Thus, the corrected 113-gram difference is surely too large; the true figure is probably close to zero and may as well favor women as men. And 113 grams, by the way, is exactly the average difference between a 5 foot 4 inch and a 6 foot 4 inch male in Broca's data. We would not (especially us short folks) want to ascribe greater intelligence to tall men. In short, who knows what to do with Broca's data? They certainly don't permit any confident claim that men have bigger brains than women.

To appreciate the social role of Broca and his school, we must recognize that his statements about the brains of women do not reflect an isolated prejudice toward a single disadvantaged group. They must be weighed in the context of a general theory that supported contemporary social distinctions as biologically ordained. Women, blacks, and poor people suffered the same disparagement, but women bore the brunt of Broca's argument because he had easier access to data on women's brains. Women were singularly denigrated but they also stood as surrogates for other disenfranchised groups. As one of Broca's disciples wrote in 1881:

"Men of the black races have a brain scarcely heavier than that of white women." This juxtaposition extended into many other realms of anthropological argument, particularly to claims that, anatomically and emotionally, both women and blacks were like white children—and that white children, by the theory of recapitulation, represented an ancestral (primitive) adult stage of human evolution. I do not regard as empty rhetoric the claim that women's battles are for all of us.

Maria Montessori did not confine her activities to educational reform for young children. She lectured on anthropology for several years at the University of Rome, and wrote an influential book entitled *Pedagogical Anthropology* (English edition, 1913). Montessori was no egalitarian. She supported most of Broca's work and the theory of innate criminality proposed by her compatriot Cesare Lombroso. She measured the circumference of children's heads in her schools and inferred that the best prospects had bigger brains. But she had no use for Broca's conclusions about women. She discussed Manouvrier's work at length and made much of his tentative claim that women, after proper correction of the data, had slightly larger brains than men. Women, she concluded, were intellectually superior, but men had prevailed heretofore by dint of physical force. Since technology has abolished force as an instrument of power, the era of women may soon be upon us: "In such an epoch there will really be superior human beings, there will really be men strong in morality and in sentiment. Perhaps in this way the reign of women is approaching, when the enigma of her anthropological superiority will be deciphered. Woman was always the custodian of human sentiment, morality and honor."

This represents one possible antidote to "scientific" claims for the constitutional inferiority of certain groups. One may affirm the validity of biological distinctions but argue that the data have been misinterpreted by prejudiced men with a stake in the outcome, and that disadvantaged groups are truly superior. In recent years, Elaine Morgan has followed this strategy in her *Descent of Woman*, a specula-

WOMEN'S BRAINS | 159

tive reconstruction of human prehistory from the woman's point of view—and as farcical as more famous tall tales by and for men.

I prefer another strategy. Montessori and Morgan followed Broca's philosophy to reach a more congenial conclusion. I would rather label the whole enterprise of setting a biological value upon groups for what it is: irrelevant and highly injurious. George Eliot well appreciated the special tragedy that biological labeling imposed upon members of disadvantaged groups. She expressed it for people like herself—women of extraordinary talent. I would apply it more widely—not only to those whose dreams are flouted but also to those who never realize that they may dream—but I cannot match her prose. In conclusion, then, the rest of Eliot's prelude to *Middlemarch:*

The limits of variation are really much wider than anyone would imagine from the sameness of women's coiffure and the favorite love stories in prose and verse. Here and there a cygnet is reared uneasily among the ducklings in the brown pond, and never finds the living stream in fellowship with its own oary-footed kind. Here and there is born a Saint Theresa, foundress of nothing, whose loving heartbeats and sobs after an unattained goodness tremble off and are dispersed among hindrances instead of centering in some long-recognizable deed.

15 | Dr. Down's Syndrome

MEIOSIS, THE SPLITTING of chromosome pairs in the formation of sex cells, represents one of the great triumphs of good engineering in biology. Sexual reproduction cannot work unless eggs and sperm each contain precisely half the genetic information of normal body cells. The union of two halves by fertilization restores the full amount of genetic information, while the mixing of genes from two parents in each offspring also supplies the variability that Darwinian processes require. This halving, or "reduction division," occurs during meiosis when the chromosomes line up in pairs and pull apart, one member of each pair moving to each of the sex cells. Our admiration for the precision of meiosis can only increase when we learn that cells of some ferns contain more than 600 pairs of chromosomes and that, in most cases, meiosis splits each pair without error.

Yet organic machines are no more infallible than their industrial counterparts. Errors in splitting often occur. On rare occasions, such errors are harbingers of new evolutionary directions. In most cases, they simply lead to misfortune for any offspring generated from the defective egg or sperm. In the most common of meiotic errors, called nondisjunction, the chromosomes fail to split. Both members of the pair go to one sex cell, while the other comes up one chromosome short. A child formed from the union of a normal sex cell with one containing an extra chromosome by nondisjunction will carry three copies of that chromosome in each cell, instead of the normal two. This anomaly is called a trisomy.

In humans, the twenty-first chromosome suffers nondisjunction at a remarkably high frequency, unfortunately rather tragic in effect. About 1 in 600 to 1 in 1,000 newborn babies carry an extra twenty-first chromosome, a condition technically known as "trisomy-21." These unfortunate children suffer mild to severe mental retardation and have a reduced life expectancy. They exhibit, in addition, a suite of distinctive features, including short and broad hands, a narrow high palate, rounded face and broad head, a small nose with a flattened root, and a thick and furrowed tongue. The frequency of trisomy-21 rises sharply with increasing maternal age. We know very little about the causes of trisomy-21; indeed, its chromosomal basis was not discovered until 1959. We have no idea why it occurs so often, and why other chromosomes are not nearly so subject to nondisjunction. We have no clue as to why an extra twenty-first chromosome should yield the highly specific set of abnormalities associated with trisomy-21. But at least it can be identified *in utero* by counting the chromosomes of fetal cells, thus providing an option for early abortion.

If this discussion strikes you as familiar, but missing in one respect, I have indeed left something out. The common designation for trisomy-21 is Mongolian idiocy, mongolism, or Down's syndrome. We have all seen children with Down's syndrome and I feel certain that I have not been alone in wondering why the condition was ever designated *Mongolian* idiocy. Most children with Down's syndrome can be recognized immediately, but (as my previous list demonstrates) their defining traits do not suggest anything oriental. Some, to be sure, have a small but perceptible epicanthic fold, the characteristic feature of an oriental eye, and some have slightly yellowish skin. These minor and inconstant features led Dr. John Langdon Haydon Down to compare them with orientals when he described the syndrome in 1866. But there is far more to the story of Down's designation than a few occasional, misleading, and superficial similarities; for it embodies an interesting tale in the history of scientific racism.

Few people who use the term are aware that both words, Mongolian and idiot, had technical meanings for Dr. Down that were rooted in the prevailing cultural prejudice, not yet extinct, for ranking people on unilinear scales with the ranker's group on top. Idiot once referred to the lowest grade in a threefold classification of mental deficiency. Idiots could never master spoken language; imbeciles, a grade above, could learn to speak but not to write. The third level, the slightly "feeble-minded," engendered considerable terminological controversy. In America, most clinicians adopted H.H. Goddard's term, "moron," from a Greek word meaning foolish. Moron is a technical term of this century, not an ancient designation, despite the length of metaphorical whiskers on those terrible, old moron jokes. Goddard, one of three major architects for the rigidly hereditarian interpretation of IQ tests, believed that his unilinear classification of mental worth could be simply extended above the level of morons to a natural ranking of human races and nationalities, with southern and eastern European immigrants on the bottom (still, on average, at moron grade), and old American WASP's on top. (After Goddard instituted IQ tests for immigrants upon their arrival at Ellis Island, he proclaimed more than 80 percent of them feeble-minded and urged their return to Europe.)

Dr. Down was medical superintendant of the Earlswood Asylum for Idiots in Surrey when he published his "Observations on an ethnic classification of idiots" in the London Hospital Reports for 1866. In a mere three pages, he managed to describe Caucasian "idiots" that reminded him of African, Malay, American Indian, and Oriental peoples. Of these fanciful comparisons, only the "idiots who arrange themselves around the Mongolian type" survived in the literature as a technical designation.

Anyone who reads Down's paper without a knowledge of its theoretical context will greatly underestimate its pervasive and serious purpose. In our perspective, it represents a set of flaky and superficial, almost whimsical, analogies presented by a prejudiced man. In his time, it embodied a deadly earnest attempt to construct a general, causal classification of mental deficiency based upon the best biological theory (and the pervasive racism) of the age. Dr. Down played for stakes higher than the identification of some curious noncausal analogies. Of previous attempts to classify mental defect, Down complained:

> Those who have given any attention to congenital mental lesions must have been frequently puzzled how to arrange, in any satisfactory way, the different classes of this defect which have come under their observation. Nor will the difficulty be lessened by an appeal to what has been written on the subject. The systems of classification are generally so vague and artificial, that, not only do they assist but feebly, in any mental arrangement of the phenomena which are presented, but they fail completely in exerting any practical influence on the subject.

In Down's day, the theory of recapitulation embodied a biologist's best guide for the organization of life into sequences of higher and lower forms. (Both the theory and "ladder approach" to classification that it encouraged are, or should be, defunct today. See my book *Ontogeny and Phylogeny*, Harvard University Press, 1977). This theory, often expressed by the mouthful "ontogeny recapitulates phylogeny," held that higher animals, in their embryonic development, pass through a series of stages representing, in proper sequence, the adult forms of ancestral, lower creatures. Thus, the human embryo first develops gill slits, like a fish, later a three chambered heart, like a reptile, still later a mammalian tail. Recapitulation provided a convenient focus for the pervasive racism of white scientists: they looked to the activities of their own children for comparison with normal, adult behavior in lower races.

As a working procedure, recapitulationists attempted to identify what Louis Agassiz had called the "threefold parallelism" of paleontology, comparative anatomy, and embryology—that is, actual ancestors in the fossil record, living representatives of primitive forms, and embryonic or youthful stages in the growth of higher animals. In the racist tradition for studying humans, the threefold parallel meant fossil ancestors (not yet discovered), "savages" or adult members of lower races, and white children.

But many recapitulationists advocated the addition of a fourth parallel—certain kinds of abnormal adults within superior races. They attributed many anomalies of form or behavior either to "throwbacks" or "arrests of development." Throwbacks, or atavisms, represent the spontaneous reappearance in adults of ancestral features that had disappeared in advanced lineages. Cesare Lombroso, for example, the founder of "criminal anthropology," believed that many lawbreakers acted by biological compulsion because a brutish past lived again in them. He sought to identify "born criminals" by "stigmata" of apish morphology—receding forehead, prominent chin, long arms.

Arrests of development represent the anomalous translation into adulthood of features that arise normally in fetal life but should be modified or replaced by something more advanced or complicated. Under the theory of recapitulation, these normal traits of fetal life are the adult stages of more primitive forms. If a Caucasian suffers developmental arrest, he may be born at a lower stage of human life—that is, he may revert to the characteristic forms of lower races. We now have a fourfold parallel of human fossil, normal adult of lower races, white children, and unfortunate white adults afflicted with atavisms or arrests of development. It is in this context that Dr. Down had his flash of fallacious insight: some Caucasian idiots must represent arrests of development and owe their mental deficiency to a retention of traits and abilities that would be judged normal in adults of lower races.

Therefore, Dr. Down scrutinized his charges for features of lower races, just as, twenty years later, Lombroso would measure the bodies of criminals for signs of apish morphology. Seek, with enough conviction aforethought, and ye shall find. Down described his search with obvious excitement: he had, or so he thought, established a natural and causal classification of mental deficiency. "I have," he wrote, "for some time had my attention directed to the possibility of making a classification of the feeble-minded, by arranging them around various ethnic standards,—in other words, framing a natural system." The more serious the deficiency, the more profound the arrest of development and the lower the race represented.

He found "several well-marked examples of the Ethiopian variety," and described their "prominent eyes," "puffy lips," and "woolly hair . . . although not always black." They are, he wrote, "specimens of white negroes, although of European descent." Next he described other idiots "that arrange themselves around the Malay variety," and still others "who with shortened foreheads, prominent cheeks, deep-set eyes, and slightly apish nose" represent those people who "originally inhabited the American continent."

Finally, mounting the scale of races, he came to the rung below Caucasian, "the great Mongolian family." "A very large number of congenital idiots," he continued, "are typical Mongols. So marked is this, that when placed side by side, it is difficult to believe that the specimens compared are not children of the same parents." Down then proceeded to describe, with fair accuracy and little indication of oriental features (beyond a "slight dirty yellowish tinge" to the skin), a boy afflicted with what we now recognize as trisomy-21, or Down's syndrome.

Down did not confine his description to supposed anatomical resemblances between oriental people and "Mongolian idiots." He also pointed to the behavior of his afflicted children: "They have considerable power of imitation, even bordering on being mimics." It requires some familiarity with the literature of nineteenth-century racism to read between these lines. The sophistication and complexity of oriental culture proved embarrassing to Caucasian racists, especially since the highest refinements of Chinese society had arisen when European culture still wallowed in barbarism. (As Benjamin Disraeli said, responding to an anti-Semitic taunt: "Yes, I am a Jew, and when the ancestors of the right honorable gentleman were brutal savages . . . mine were priests in the temple of Solomon.") Caucasians solved this dilemma by admitting the intellectual power of orientals, but attributing it to a facility for imitative copying, rather than to innovative genius.

Down concluded his description of a child with trisomy-21 by attributing the condition to developmental arrest (due, Down thought, to the tubercular condition of his parents): "The boy's aspect is such that it is difficult to realize that he is the child of Europeans, but so frequently are these characters presented, that there can be no doubt that these ethnic features are the result of degeneration."

By the standards of his time, Down was something of a racial "liberal." He argued that all people had descended from the same stock and could be united into a single family, with gradation by status to be sure. He used his ethnic classification of idiots to combat the claim of some scientists that lower races represented separate acts of creation and could not "improve" towards whiteness. He wrote:

> If these great racial divisions are fixed and definite, how comes it that disease is able to break down the barrier, and to simulate so closely the features of the members of another division. I cannot but think that the observations which I have recorded, are indications that the differences in the races are not specific but variable. These examples of the result of degeneracy among mankind, appear to me to furnish some arguments in favor of the unity of the human species.

Down's general theory of mental deficiency enjoyed some popularity, but never swept the field. Yet his name for one specific anomaly, Mongolian idiocy (sometimes softened to mongolism) stuck long after most physicians forgot why Down had coined the term. Down's own son rejected his father's comparison of orientals and children with trisomy-21, though he defended both the low status of orientals and the general theory linking mental deficiency with evolutionary reversion:

> It would appear that the characteristics which at first sight strikingly suggest Mongolian features and build are accidental and superficial, being constantly associated, as they are, with other features which are in no way characteristic of that race, and if this is a case of reversion it must be reversion to a type even further back than the Mongol stock, from which some ethnologists believe all the various races of men have sprung.

Down's theory for trisomy-21 lost its rationale—even within Down's invalid racist system—when physicians detected it both in orientals themselves, and in races lower than oriental by Down's classification. (One physician referred to "Mongol Mongolians" but that clumsy perseverance never took hold.) The condition could scarcely be due to degeneration if it represented the normal state of a higher race. We now know that a similar set of features occurs in some chimpanzees who carry an extra chromosome probably homologous with the twenty-first of humans.

With Down's theory disproved, what should become of his term? A few years ago, Sir Peter Medawar and a group of oriental scientists persuaded several British publications to substitute Down's syndrome for Mongolian idiocy and mongolism. I detect a similar trend in this country, although mongolism is still commonly used. Some people may complain that efforts to change the name represent yet another misguided attempt by fuzzy-minded liberals to muck around with accepted usage by introducing social concerns into realms where they don't belong. Indeed, I do not believe in capricious alteration of established names. I suffer extreme discomfort every time I sing in Bach's St. Matthew Passion and must, as an angry member of the Jewish crowd, shout out the passage that served for centuries as an "official" justification for anti-Semitism: Sein Blut komme über uns und unsre Kinder—"His blood be upon us and upon our children." Yet, as he to whom the passage refers said in another context, I would not change "one jot or one tittle" of Bach's text.

But scientific names are not literary monuments. Mongolian idiocy is not only defamatory. It is wrong on all counts. We no longer classify mental deficiency as a unilinear sequence. Children with Down's syndrome do not resemble orientals to any great extent, if at all. And, most importantly, the name only has meaning in the context of Down's discredited theory of racial reversion as the cause of mental deficiency. If we must honor the good doctor, then let his name stand as a designation for trisomy-21—Down's syndrome.

16 | Flaws in a Victorian Veil

THE VICTORIANS LEFT some magnificent, if lengthy, novels. But they also foisted upon an apparently willing world a literary genre probably unmatched for tedium and inaccurate portrayal: the multivolumed "life and letters" of eminent men. These extended encomiums, usually written by grieving widows or dutiful sons and daughters, masqueraded as humbly objective accounts, simple documentation of words and activities. If we accepted these works at face value, we would have to believe that eminent Victorians actually lived by the ethical values they espoused—a fanciful proposition that Lytton Strachey's *Eminent Victorians* put to rest more than fifty years ago.

Elizabeth Cary Agassiz—eminent Bostonian, founder and first president of Radcliffe College, and devoted wife of America's premier naturalist—had all the right credentials for authorship (including a departed and lamented husband). Her *Life and Correspondence of Louis Agassiz* turned a fascinating, cantankerous, and not overly faithful man into a paragon of restraint, statesmanship, wisdom, and rectitude.

I write this essay in the structure that Louis Agassiz built in 1859—the original wing of Harvard's Museum of Comparative Zoology. Agassiz, the world's leading student of fossil fishes, protégé of the great Cuvier (see essay 13), left his native Switzerland for an American career in the late

1840s. As a celebrated European and a charming man, Agassiz was lionized in social and intellectual circles from Boston to Charleston. He led the study of natural history in America until his death in 1873.

Louis's public utterances were always models of propriety, but I expected that his private letters would match his ebullient personality. Yet Elizabeth's book, ostensibly a verbatim report of Louis's letters, manages to turn this focus of controversy and source of restless energy into a measured and dignified gentleman.

Recently, in studying Louis Agassiz's views on race and prompted by some hints in E. Lurie's biography *(Louis Agassiz: a life in science)*, I encountered some interesting discrepancies between Elizabeth's version and Louis's original letters. I then discovered that Elizabeth simply expurgated the text and didn't even insert ellipses (those annoying three dots) to indicate her deletions. Harvard has the original letters, and a bit of sleuthing on my part turned up some spicy material.

During the decade before the Civil War, Agassiz expressed strong opinions on the status of blacks and Indians. As an adopted son of the north, he rejected slavery, but as an upper crust Caucasian, he certainly didn't link this rejection to any notion of racial equality.

Agassiz presented his racial attitudes as sober and ineluctable deductions from first principles. He maintained that species are static, created entities (at his death in 1873, Agassiz stood virtually alone among biologists as a holdout against the Darwinian tide). They are not placed upon the earth in a single spot, but created simultaneously over their entire range. Related species are often created in separate geographic regions, each adapted to prevailing environments of its own area. Since human races met these criteria before commerce and migration mixed us up, each race is a separate biological species.

Thus, America's leading biologist came down firmly on the wrong side of a debate that had been raging in America for a decade before he arrived: Was Adam the progenitor of all people or only of white people? Are blacks and Indians our brothers or merely our look-alikes? The *polygenists*, Agassiz among them, held that each major race had been created as a truly separate species; the *monogenists* advocated a single origin and ranked races by their unequal degeneration from the primeval perfection of Eden—the debate included no egalitarians. In logic, separate needn't mean unequal, as the victors in Plessy vs. Ferguson argued in 1896. But, as the winners in Brown vs. the Topeka Board of Education maintained in 1954, a group in power always conflates separation with superiority. I know of no American polygenist who did not assume that whites were separate *and* superior.

Agassiz insisted that his defense of polygeny had nothing to do with political advocacy or social prejudice. He was, he argued, merely a humble and disinterested scholar, trying to establish an intriguing fact of natural history.

> It has been charged upon the views here advanced that they tend to the support of slavery. . . . Is that a fair objection to a philosophical investigation? Here we have to do only with the question of the origin of men; let the politicians, let those who feel themselves called upon to regulate human society, see what they can do with the results. . . . We disclaim all connection with any question involving political matters. . . . Naturalists have a right to consider the questions growing out of men's physical relations as merely scientific questions, and to investigate them without reference to either politics or religion.

Despite these brave words, Agassiz ends this major statement on race (published in the *Christian Examiner*, 1850) with some definite social recommendations. He begins by affirming the doctrine of separate and unequal: "There are upon earth different races of men, inhabiting different parts of its surface . . . and this fact presses upon us the obligation to settle the relative rank among these races." The resulting hierarchy is self-evident: "The indomitable, courageous, proud Indian—in how different a light he stands by the side of the submissive, obsequious, imitative negro, or by the side of the tricky, cunning, and cowardly Mongolian! Are not these facts indications that the different races do not rank upon one level in nature." Finally, if he hadn't made his political message clear by generalization, Agassiz ends by advocating specific social policy—thus contravening his original pledge to abjure politics for the pure life of the mind. Education, he argues, must be tailored to innate ability; train blacks in hand work, whites in mind work.

> What would be the best education to be imparted to the different races in consequence of their primitive difference. . . . We entertain not the slightest doubt that human affairs with reference to the colored races would be far more judiciously conducted if, in our intercourse with them, we were guided by a full consciousness of the real differences existing between us and them, and a desire to foster those dispositions that are eminently marked in them, rather than by treating them on terms of equality.

Since these "eminently marked" dispositions are submissiveness, obsequiousness, and imitation, we can well imagine what Agassiz had in mind.

Agassiz had political clout, largely because he spoke as a scientist, supposedly motivated only by the facts of his case and the abstract theory they embodied. In this context, the actual source of Agassiz's ideas on race becomes a matter of some importance. Did he really have no ax to grind, no predisposition, no impetus beyond his love for natural history? The passages expurgated from *Life and Correspondence* shed considerable light. They show a man with strong prejudices based primarily on immediate visceral reactions and deep sexual fears.

The first passage, almost shocking in its force, even 130 years later, recounts Agassiz's first experience with black people (he had never encountered blacks in Europe). He first visited America in 1846 and sent his mother a long letter detailing his experiences. In the section about Philadelphia, Elizabeth Agassiz records only his visits to museums and the private homes of scientists. She expunges, without ellipses, his first impression of blacks—a visceral reaction to waiters in a hotel restaurant. In 1846 Agassiz still believed in human unity, but this passage exposes an explicit, stunningly nonscientific basis for his conversion to polygeny. For the first time, then, without omissions:

> It was in Philadelphia that I first found myself in prolonged contact with negroes; all the domestics in my hotel were men of color. I can scarcely express to you the painful impression that I received, especially since the sentiment that they inspired in me is contrary to all our ideas about the confraternity of the human type and the unique origin of our species. But truth before all. Nevertheless, I experienced pity at the sight of this degraded and degenerate race, and their lot inspired compassion in me in thinking that they are really men. Nonetheless, it is impossible for me to repress the feeling that they are not of the same blood as us. In seeing their black faces with their thick lips and grimacing teeth, the wool on their head, their bent knees, their elongated hands, their large curved nails, and especially the livid color of the palms of their hands, I could not take my eyes off their faces in order to tell them to stay far away. And when they advanced that hideous hand towards my plate in order to serve me, I wished I were able to depart in order to eat a piece of bread elsewhere, rather than to dine with such service. What unhappiness for the white race—to have tied their existence so closely with that of negroes in certain countries! God preserve us from such a contact!

The second set of documents comes from the midst of the Civil War. Samuel Howe, husband of Julia Ward Howe (author of *The Battle Hymn of the Republic*) and a member of President Lincoln's Inquiry Commission, wrote to ask Agassiz his opinion about the role of blacks in a reunited nation. During August 1863, Agassiz responded in four long and impassioned letters. Elizabeth Agassiz bowdlerized them to render Louis's case as a soberly stated opinion (despite its peculiar content), derived from first principles and motivated only by a love of truth.

Louis argued, in short, that races should be kept separate lest white superiority be diluted. This separation should occur naturally since mulattoes, as a weak strain, will eventually die out. Blacks will leave the northern climates so unsuited to them (since they were created as a separate species for Africa); they will move south in droves and will eventually prevail in a few lowland states, although whites will maintain dominion over the seashore and elevated ground. We will have to recognize these states, even admit them to the Union, as the best solution to a bad situation; after all, we do recognize "Haity and Liberia."

Elizabeth's substantial deletions display Louis's motivation in a very different light. They radiate raw fear and blind prejudice. She systematically eliminates three kinds of statements. First, she omits the most denigrating references to blacks: "In everything unlike other races," Louis writes, "they may but be compared to children, grown in the stature of adults, while retaining a childlike mind." Second, she removes all elitist claims about the correlation of wisdom, wealth, and social position within races. In these passages, we begin to sense Louis's real fears about miscegenation.

I shudder from the consequences. We have already to struggle, in our progress, against the influence of universal equality, in consequence of the difficulty of preserving the acquisitions of individual eminence, the wealth of refinement and culture growing out of select associations. What would be our condition if to these difficulties were added the far more tenacious influences of physical disability. Improvements in our system of education . . . may sooner or later counterbalance the effects of the apathy of the uncultivated and of the rudeness of the lower classes and raise them to a higher standard. But how shall we eradicate the stigma of a lower race when its blood has once been allowed to flow freely into that of our children.

Third, and of greatest significance, she expunges several long passages on interbreeding that place the entire correspondence in a different setting from the one she fashioned. In them, we grasp Louis's intense, visceral revulsion toward the idea of sexual contact between races. This deep and irrational fear was as strong a driving force within him as any abstract notion about separate creation: "The production of half-breeds," he writes, "is as much a sin against nature, as incest in a civilized community is a sin against purity of character. . . . I hold it to be a perversion of every natural sentiment."

This natural aversion is so strong that abolitionist sentiment cannot reflect any innate sympathy for blacks but must arise because many "blacks" have substantial amounts of white blood and whites instinctively sense this part of themselves: "I have no doubt in my mind that the sense of abhorrence against slavery, which has led to the agitation now culminating in our civil war, has been chiefly if unconsciously fostered by the recognition of our own type in the offspring of southern gentlemen, moving among us as negros [sic], which they are not."

But if races naturally repel each other, how then do "southern gentlemen" take such willing advantage of their bonded women? Agassiz blames the mulatto house slaves. Their whiteness renders them attractive; their blackness, lascivious. The poor, innocent young men are enticed and entrapped.

As soon as the sexual desires are awakening in the young men of the South, they find it easy to gratify themselves by the readiness with which they are met by colored [mulatto] house servants. [This contact] blunts his better instincts in that direction and leads him gradually to seek more spicy partners, as I have heard the full blacks called by fast young men. One thing is certain, that there is no elevating element whatever conceivable in the connection of individuals of different races; there is neither love, nor desire for improvement of any kind. It is altogether a physical connection.

How a previous generation of gentlemen overcame their aversion to produce the first mulattoes, we are not told.

We cannot know in detail why Elizabeth chose her deletions. I doubt that a conscious desire to convert Louis's motives from prejudice to logical implication prompted all her actions. Simple Victorian prudery probably led her to reject a public airing of any statement about sex. In any case, her deletions did distort Louis Agassiz's thought and did render his intentions according to the fallacious and self-serving model favored by scientists—that opinions arise from dispassionate surveys of raw information.

These restorations show that Louis Agassiz was jolted to consider the polygenist theory of races as separate species by his initial, visceral reaction to contact with blacks. They also demonstrate that his extreme views on racial mixing were powered more by intense sexual revulsion than by any abstract theory of hybridity.

Racism has often been buttressed by scientists who present a public façade of objectivity to mask their guiding prejudices. Agassiz's case may be distant, but its message rings through our century as well.

5 | The Pace of Change

17 | The Episodic Nature of Evolutionary Change

ON NOVEMBER 23, 1859, the day before his revolutionary book hit the stands, Charles Darwin received an extraordinary letter from his friend Thomas Henry Huxley. It offered warm support in the coming conflict, even the supreme sacrifice: "I am prepared to go to the stake, if requisite . . . I am sharpening up my claws and beak in readiness." But it also contained a warning: "You have loaded yourself with an unnecessary difficulty in adopting *Natura non facit saltum* so unreservedly."

The Latin phrase, usually attributed to Linnaeus, states that "nature does not make leaps." Darwin was a strict adherent to this ancient motto. As a disciple of Charles Lyell, the apostle of gradualism in geology, Darwin portrayed evolution as a stately and orderly process, working at a speed so slow that no person could hope to observe it in a lifetime. Ancestors and descendants, Darwin argued, must be connected by "infinitely numerous transitional links" forming "the finest graduated steps." Only an immense span of time had permitted such a sluggish process to achieve so much.

Huxley felt that Darwin was digging a ditch for his own theory. Natural selection required no postulate about rates; it could operate just as well if evolution proceeded at a rapid pace. The road ahead was rocky enough; why harness the theory of natural selection to an assumption both unnecessary and probably false? The fossil record offered no support for gradual change: whole faunas had been wiped out during disarmingly short intervals. New species almost always appeared suddenly in the fossil record with no intermediate links to ancestors in older rocks of the same region. Evolution, Huxley believed, could proceed so rapidly that the slow and fitful process of sedimentation rarely caught it in the act.

The conflict between adherents of rapid and gradual change had been particularly intense in geological circles during the years of Darwin's apprenticeship in science. I do not know why Darwin chose to follow Lyell and the gradualists so strictly, but I am certain of one thing: preference for one view or the other had nothing to do with superior perception of empirical information. On this question, nature spoke (and continues to speak) in multifarious and muffled voices. Cultural and methodological preferences had as much influence upon any decision as the constraints of data.

On issues so fundamental as a general philosophy of change, science and society usually work hand in hand. The static systems of European monarchies won support from legions of scholars as the embodiment of natural law. Alexander Pope wrote:

Order is Heaven's first law; and this confessed,
Some are, and must be, greater than the rest.

As monarchies fell and as the eighteenth century ended in an age of revolution, scientists began to see change as a normal part of universal order, not as aberrant and exceptional. Scholars then transferred to nature the liberal program of slow and orderly change that they advocated for social transformation in human society. To many scientists, natural cataclysm seemed as threatening as the reign of terror that had taken their great colleague Lavoisier.

Yet the geologic record seemed to provide as much evidence for cataclysmic as for gradual change. Therefore, in defending gradualism as a nearly universal tempo, Darwin had to use Lyell's most characteristic method of argument

—he had to reject literal appearance and common sense for an underlying "reality." (Contrary to popular myths, Darwin and Lyell were not the heros of true science, defending objectivity against the theological fantasies of such "catastrophists" as Cuvier and Buckland. Catastrophists were as committed to science as any gradualist; in fact, they adopted the more "objective" view that one should believe what one sees and not interpolate missing bits of a gradual record into a literal tale of rapid change.) In short, Darwin argued that the geologic record was exceedingly imperfect—a book with few remaining pages, few lines on each page, and few words on each line. We do not see slow evolutionary change in the fossil record because we study only one step in thousands. Change seems to be abrupt because the intermediate steps are missing.

The extreme rarity of transitional forms in the fossil record persists as the trade secret of paleontology. The evolutionary trees that adorn our textbooks have data only at the tips and nodes of their branches; the rest is inference, however reasonable, not the evidence of fossils. Yet Darwin was so wedded to gradualism that he wagered his entire theory on a denial of this literal record:

> The geological record is extremely imperfect and this fact will to a large extent explain why we do not find interminable varieties, connecting together all the extinct and existing forms of life by the finest graduated steps. He who rejects these views on the nature of the geological record, will rightly reject my whole theory.

Darwin's argument still persists as the favored escape of most paleontologists from the embarrassment of a record that seems to show so little of evolution directly. In exposing its cultural and methodological roots, I wish in no way to impugn the potential validity of gradualism (for all general views have similar roots). I wish only to point out that it was never "seen" in the rocks.

Paleontologists have paid an exorbitant price for Darwin's argument. We fancy ourselves as the only true students of life's history, yet to preserve our favored account of evolution by natural selection we view our data as so bad that we almost never see the very process we profess to study.

For several years, Niles Eldredge of the American Museum of Natural History and I have been advocating a resolution of this uncomfortable paradox. We believe that Huxley was right in his warning. The modern theory of evolution does not require gradual change. In fact, the operation of Darwinian processes should yield exactly what we see in the fossil record. It is gradualism that we must reject, not Darwinism.

The history of most fossil species includes two features particularly inconsistent with gradualism:

1. *Stasis.* Most species exhibit no directional change during their tenure on earth. They appear in the fossil record looking much the same as when they disappear; morphological change is usually limited and directionless.

2. *Sudden appearance.* In any local area, a species does not arise gradually by the steady transformation of its ancestors; it appears all at once and "fully formed."

Evolution proceeds in two major modes. In the first, phyletic transformation, an entire population changes from one state to another. If all evolutionary change occurred in this mode, life would not persist for long. Phyletic evolution yields no increase in diversity, only a transformation of one thing into another. Since extinction (by extirpation, not by evolution into something else) is so common, a biota with no mechanism for increasing diversity would soon be wiped out. The second mode, speciation, replenishes the earth. New species branch off from a persisting parental stock.

Darwin, to be sure, acknowledged and discussed the process of speciation. But he cast his discussion of evolutionary change almost totally in the mold of phyletic transformation. In this context, the phenomena of stasis and sudden appearance could hardly be attributed to anything but imperfection of the record; for if new species arise by the transformation of entire ancestral populations, and if we almost never see the transformation (because species are essentially static through their range), then our record must be hopelessly incomplete.

Eldredge and I believe that speciation is responsible for almost all evolutionary change. Moreover, the way in which it occurs virtually guarantees that sudden appearance and stasis shall dominate the fossil record.

All major theories of speciation maintain that splitting takes place rapidly in very small populations. The theory of geographic, or allopatric, speciation is preferred by most evolutionists for most situations (allopatric means "in another place").* A new species can arise when a small segment of the ancestral population is isolated at the periphery of the ancestral range. Large, stable central populations exert a strong homogenizing influence. New and favorable mutations are diluted by the sheer bulk of the population through which they must spread. They may build slowly in frequency, but changing environments usually cancel their selective value long before they reach fixation. Thus, phyletic transformation in large populations should be very rare—as the fossil record proclaims.

But small, peripherally isolated groups are cut off from their parental stock. They live as tiny populations in geographic corners of the ancestral range. Selective pressures

---

*I wrote this essay in 1977. Since then, a major shift of opinion has been sweeping through evolutionary biology. The allopatric orthodoxy has been breaking down and several mechanisms of sympatric speciation have been gaining both legitimacy and examples. (In sympatric speciation, new forms arise within the geographic range of their ancestors.) These sympatric mechanisms are united in their insistence upon the two conditions that Eldredge and I require for our model of the fossil record—*rapid* origin in a *small* population. In fact, they generally advocate smaller groups and more rapid change than conventional allopatry envisages (primarily because groups in potential contact with their forebears must move quickly towards reproductive isolation, lest their favorable variants be diluted by breeding with the more numerous parental forms). See White (1978) for a thorough discussion of these sympatric models.

are usually intense because peripheries mark the edge of ecological tolerance for ancestral forms. Favorable variations spread quickly. Small, peripheral isolates are a laboratory of evolutionary change.

What should the fossil record include if most evolution occurs by speciation in peripheral isolates? Species should be static through their range because our fossils are the remains of large central populations. In any local area inhabited by ancestors, a descendent species should appear suddenly by migration from the peripheral region in which it evolved. In the peripheral region itself, we might find direct evidence of speciation, but such good fortune would be rare indeed because the event occurs so rapidly in such a small population. Thus, the fossil record is a faithful rendering of what evolutionary theory predicts, not a pitiful vestige of a once bountiful tale.

Eldredge and I refer to this scheme as the model of *punctuated equilibria.* Lineages change little during most of their history, but events of rapid speciation occasionally punctuate this tranquillity. Evolution is the differential survival and deployment of these punctuations. (In describing the speciation of peripheral isolates as very rapid, I speak as a geologist. The process may take hundreds, even thousands of years; you might see nothing if you stared at speciating bees on a tree for your entire lifetime. But a thousand years is a tiny fraction of one percent of the average duration for most fossil invertebrate species—5 to 10 million years. Geologists can rarely resolve so short an interval at all; we tend to treat it as a moment.)

If gradualism is more a product of Western thought than a fact of nature, then we should consider alternate philosophies of change to enlarge our realm of constraining prejudices. In the Soviet Union, for example, scientists are trained with a very different philosophy of change—the so-called dialectical laws, reformulated by Engels from Hegel's philosophy. The dialectical laws are explicitly punctuational. They speak, for example, of the "transformation of quantity into quality." This may sound like mumbo jumbo, but it suggests that change occurs in large leaps following a slow accumulation of stresses that a system resists until it reaches the breaking point. Heat water and it eventually boils. Oppress the workers more and more and bring on the revolution. Eldredge and I were fascinated to learn that many Russian paleontologists support a model similar to our punctuated equilibria.

I emphatically do not assert the general "truth" of this philosophy of punctuational change. Any attempt to support the exclusive validity of such a grandiose notion would border on the nonsensical. Gradualism sometimes works well. (I often fly over the folded Appalachians and marvel at the striking parallel ridges left standing by gradual erosion of the softer rocks surrounding them.) I make a simple plea for pluralism in guiding philosophies, and for the recognition that such philosophies, however hidden and unarticulated, constrain all our thought. The dialectical laws express an ideology quite openly; our Western preference for gradualism does the same thing more subtly.

Nonetheless, I will confess to a personal belief that a punctuational view may prove to map tempos of biological and geologic change more accurately and more often than any of its competitors—if only because complex systems in steady state are both common and highly resistant to change. As my colleague British geologist Derek V. Ager writes in supporting a punctuational view of geologic change: "The history of any one part of the earth, like the life of a soldier, consists of long periods of boredom and short periods of terror."

18 | Return of the Hopeful Monster

BIG BROTHER, THE tyrant of George Orwell's *1984,* directed his daily Two Minutes Hate against Emmanuel Goldstein, enemy of the people. When I studied evolutionary biology in graduate school during the mid-1960s, official rebuke and derision focused upon Richard Goldschmidt, a famous geneticist who, we were told, had gone astray. Although 1984 creeps up on us, I trust that the world will not be in Big Brother's grip by then. I do, however, predict that during this decade Goldschmidt will be largely vindicated in the world of evolutionary biology.

Goldschmidt, a Jewish refugee from Hitler's decimation of German science, spent the remainder of his career at Berkeley, where he died in 1958. His views on evolution ran afoul of the great neo-Darwinian synthesis forged during the 1930s and 1940s and continuing today as a reigning, if insecure, orthodoxy. Contemporary neo-Darwinism is often called the "synthetic theory of evolution" because it united the theories of population genetics with the classical observations of morphology, systematics, embryology, biogeography, and paleontology.

The core of this synthetic theory restates the two most characteristic assertions of Darwin himself: first, that evolution is a two-stage process (random variation as raw material, natural selection as a directing force); secondly, that evolutionary change is generally slow, steady, gradual, and continuous.

Geneticists can study the gradual increase of favored genes within populations of fruit flies in laboratory bottles. Naturalists can record the steady replacement of light moths by dark moths as industrial soot blackens the trees of Britain. Orthodox neo-Darwinians extrapolate these even and continuous changes to the most profound structural transitions in the history of life: by a long series of insensibly graded intermediate steps, birds are linked to reptiles, fish with jaws to their jawless ancestors. Macroevolution (major structural transition) is nothing more than microevolution (flies in bottles) extended. If black moths can displace white moths in a century, then reptiles can become birds in a few million years by the smooth and sequential summation of countless changes. The shift of gene frequencies in local populations is an adequate model for all evolutionary processes—or so the current orthodoxy states.

The most sophisticated of modern American textbooks for introductory biology expresses its allegiance to the conventional view in this way:

> [Can] more extensive evolutionary change, macroevolution, be explained as an outcome of these microevolutionary shifts? Did birds really arise from reptiles by an accumulation of gene substitutions of the kind illustrated by the raspberry eye-color gene?
>
> The answer is that it is entirely plausible, and no one has come up with a better explanation. . . . The fossil record suggests that macroevolution is indeed gradual, paced at a rate that leads to the conclusion that it is based upon hundreds or thousands of gene substitutions no different in kind from the ones examined in our case histories.

Many evolutionists view strict continuity between micro- and macroevolution as an essential ingredient of Darwinism and a necessary corollary of natural selection. Yet, as I argue in essay 17, Thomas Henry Huxley divided the two issues of natural selection and gradualism and warned Darwin that his strict and unwarranted adherence to gradualism might undermine his entire system. The fossil record with its abrupt transitions offers no support for gradual change, and the principle of natural selection does not require it— selection can operate rapidly. Yet the unnecessary link that Darwin forged became a central tenet of the synthetic theory.

Goldschmidt raised no objection to the standard accounts of microevolution; he devoted the first half of his major work, *The Material Basis of Evolution* (Yale University Press, 1940), to gradual and continuous change within species. He broke sharply with the synthetic theory, however, in arguing that new species arise abruptly by discontinuous variation, or macromutation. He admitted that the vast majority of macromutations could only be viewed as disastrous —these he called "monsters." But, Goldschmidt continued, every once in a while a macromutation might, by sheer good fortune, adapt an organism to a new mode of life, a "hopeful monster" in his terminology. Macroevolution proceeds by the rare success of these hopeful monsters, not by an accumulation of small changes within populations.

I want to argue that defenders of the synthetic theory made a caricature of Goldschmidt's ideas in establishing their whipping boy. I shall not defend everything Goldschmidt said; indeed, I disagree fundamentally with his claim that abrupt macroevolution discredits Darwinism. For Goldschmidt also failed to heed Huxley's warning that the essence of Darwinism—the control of evolution by natural selection—does not require a belief in gradual change.

As a Darwinian, I wish to defend Goldschmidt's postulate that macroevolution is not simply microevolution extrapolated, and that major structural transitions can occur rapidly without a smooth series of intermediate stages. I shall proceed by discussing three questions: (1) can a reasonable story of continuous change be constructed for all macroevolutionary events? (my answer shall be no); (2) are theories of abrupt change inherently anti-Darwinian? (I shall argue that some are and some aren't); (3) do Goldschmidt's hopeful monsters represent the archetype of apostasy from Darwinism, as his critics have long maintained? (my answer, again, shall be no).

All paleontologists know that the fossil record contains precious little in the way of intermediate forms; transitions between major groups are characteristically abrupt. Gradualists usually extract themselves from this dilemma by invoking the extreme imperfection of the fossil record—if only one step in a thousand survives as a fossil, geology will not record continuous change. Although I reject this argument (for reasons discussed in essay 17), let us grant the traditional escape and ask a different question. Even though we have no direct evidence for smooth transitions, can we invent a reasonable sequence of intermediate forms—that is, viable, functioning organisms—between ancestors and descendants in major structural transitions? Of what possible use are the imperfect incipient stages of useful structures? What good is half a jaw or half a wing? The concept of *preadaptation* provides the conventional answer by permitting us to argue that incipient stages performed different functions. The half jaw worked perfectly well as a series of gill-supporting bones; the half wing may have trapped prey or controlled body temperature. I regard preadaptation as an important, even an indispensable, concept. But a plausible story is not necessarily true. I do not doubt that preadaptation can save gradualism in some cases, but does it permit us to invent a tale of continuity in most or all cases? I submit, although it may only reflect my lack of imagination, that the answer is no, and I invoke two recently supported cases of discontinuous change in my defense.

On the isolated island of Mauritius, former home of the dodo, two genera of boid snakes (a large group that includes pythons and boa constrictors) share a feature present in no other terrestrial vertebrate: the maxillary bone of the upper jaw is split into front and rear halves, connected by a movable joint. In 1970, my friend Tom Frazzetta published a paper entitled "From Hopeful Monsters to Bolyerine Snakes?" He considered every preadaptive possibility he could imagine and rejected them in favor of discontinuous transition. How can a jawbone be half broken?

Many rodents have cheek pouches for storing food. These internal pouches connect to the pharynx and may have evolved gradually under selective pressure for holding more and more food in the mouth. But the Geomyidae (pocket gophers) and Heteromyidae (kangaroo rats and pocket mice) have invaginated their cheeks to form external fur-lined pouches with no connection to the mouth or pharynx. What good is an incipient groove or furrow on the outside? Did such hypothetical ancestors run about three-legged while holding a few scraps of food in an imperfect crease with their fourth leg? Charles A. Long has recently considered a suite of preadaptive possibilities (external grooves in burrowing animals to transport soil, for example) and rejected them all in favor of discontinuous transition. These tales, in the "just-so story" tradition of evolutionary natural history, do not prove anything. But the weight of these, and many similar cases, wore down my faith in gradualism long ago. More inventive minds may yet save it, but concepts salvaged only by facile speculation do not appeal much to me.

If we must accept many cases of discontinuous transition in macroevolution, does Darwinism collapse to survive only as a theory of minor adaptive change within species? The essence of Darwinism lies in a single phrase: natural selection is the major creative force of evolutionary change. No one denies that natural selection will play a negative role in eliminating the unfit. Darwinian theories require that it create the fit as well. Selection must do this by building adaptations in a series of steps, preserving at each stage the advantageous part in a random spectrum of genetic variability. Selection must superintend the process of creation, not just toss out the misfits after some other force suddenly produces a new species, fully formed in pristine perfection.

We can well imagine such a non-Darwinian theory of discontinuous change—profound and abrupt genetic alteration luckily (now and then) making a new species all at once. Hugo de Vries, the famous Dutch botanist, supported such a theory early in this century. But these notions seem to present insuperable difficulties. With whom shall Athena born from Zeus's brow mate? All her relatives are members of another species. What is the chance, of producing Athena in the first place, rather than a deformed monster? Major disruptions of entire genetic systems do not produce favored—or even viable—creatures.

But all theories of discontinuous change are not anti-Darwinian, as Huxley pointed out nearly 120 years ago. Suppose that a discontinuous change in adult form arises from a small genetic alteration. Problems of discordance with other members of the species do not arise, and the large, favorable variant can spread through a population in Darwinian fashion. Suppose also that this large change does not produce a perfected form all at once, but rather serves as a "key" adaptation to shift its possessor toward a new mode of life. Continued success in this new mode may require a large set of collateral alterations, morphological and behavioral; these may arise by a more traditional, gradual route once the key adaptation forces a profound shift in selective pressures.

Defenders of the modern synthesis have cast Goldschmidt as Goldstein by linking his catchy phrase—hopeful monster—to non-Darwinian notions of immediate perfection by profound genetic change. But this is not entirely what Goldschmidt maintained. In fact, one of his mechanisms for discontinuity in adult forms relied upon a notion of small underlying genetic change. Goldschmidt was a student of embryonic development. He spent most of his early career studying geographic variation in the gypsy moth, *Lymantria dispar.* He found that large differences in the color patterns of caterpillars resulted from small changes in the timing of development: the effects of a slight delay or enhancement of pigmentation early in growth increased through ontogeny and led to profound differences among fully grown caterpillars.

Goldschmidt identified the genes responsible for these small changes in timing, and demonstrated that large final differences reflected the action of one or a few "rate genes" acting early in growth. He codified the notion of a rate gene in 1918 and wrote twenty years later:

The mutant gene produces its effect . . . by changing the rates of partial processes of development. These might be rates of growth or differentiation, rates of production of stuffs necessary for differentiation, rates of reactions leading to definite physical or chemical situations at definite times of development, rates of those processes which are responsible for segregating the embryonic potencies at definite times.

In his infamous book of 1940, Goldschmidt specifically invokes rate genes as a potential maker of hopeful monsters: "This basis is furnished by the existence of mutants producing monstrosities of the required type and the knowledge of embryonic determination, which permits a small rate change in early embryonic processes to produce a large effect embodying considerable parts of the organism."

In my own, strongly biased opinion, the problem of reconciling evident discontinuity in macroevolution with Darwinism is largely solved by the observation that small changes early in embryology accumulate through growth to yield profound differences among adults. Prolong the high prenatal rate of brain growth into early childhood and a monkey's brain moves toward human size. Delay the onset of metamorphosis and the axolotl of Lake Xochimilco reproduces as a tadpole with gills and never transforms into a salamander. (See my book *Ontogeny and Phylogeny* [Harvard University Press, 1977] for a compendium of examples, and pardon me for the unabashed plug.) As Long argues for the external cheek pouch: "A genetically controlled developmental inversion of the cheek pouch may have occurred, recurred, and persisted in some populations. Such a morphological change would have been drastic in effect, turning the pockets 'wrong side out' (furry side in), but nevertheless it would be a rather simple embryonic change."

Indeed, if we do not invoke discontinuous change by small alteration in rates of development, I do not see how most major evolutionary transitions can be accomplished at all. Few systems are more resistant to basic change than the strongly differentiated, highly specified, complex adults of "higher" animal groups. How could we ever convert an adult rhinoceros or a mosquito into something fundamentally different. Yet transitions between major groups have occurred in the history of life.

D'Arcy Wentworth Thompson, classical scholar, Victorian prose stylist, and glorious anachronism of twentieth-century biology, dealt with this dilemma in his classic treatise *On Growth and Form.*

> An algebraic curve has its fundamental formula, which defines the family to which it belongs. . . . We never think of "transforming" a helicoid into an ellipsoid, or a circle into a frequency curve. So it is with the forms of animals. We cannot transform an invertebrate into a vertebrate, nor a coelenterate into a worm, by any simple and legitimate deformation. . . . Nature proceeds from one type to another. . . . To seek for steppingstones across the gaps between is to seek in vain, forever.

D'Arcy Thompson's solution was the same as Goldschmidt's: the transition may occur in simpler and more similar embryos of these highly divergent adults. No one would think of transforming a starfish into a mouse, but the embryos of some echinoderms and protovertebrates are nearly identical.

1984 will mark the 125th anniversary of Darwin's *Origin,* the first major excuse for a celebration since the centenary of 1959. I hope that our "new speaking" these few years hence will be neither dogma nor vacuous nonsense. If our entrenched, a priori preferences for gradualism begin to fade by then, we may finally be able to welcome the plurality of results that nature's complexity provides.

19 | The Great Scablands Debate

THE INTRODUCTORY PARAGRAPHS
of popular guidebooks usually tout prevailing orthodoxy in
its purest form—dogma unadulterated by the "howevers"
of professional writing. Consider the following from our
National Park Service's auto tour of Arches National Park:

> The world and all it contains is in a continuous pro-
> cess of change. Most of the changes in our world are
> very tiny and so escape our notice. They are real, how-
> ever, and over an immense span of time their combined
> effect is to bring about great change. If you stand at the
> base of a canyon wall and rub your hand on the sand-
> stone, hundreds of grains of sand are dislodged. It
> seems like an insignificant change, but that's how the
> canyon was formed. Various forces have dislodged and
> carried away grains of sand. Sometimes the process is
> "very fast" (as when you rub the sandstone) but most
> of the time it is much slower. If you allow sufficient
> time, you can tear down a mountain or create a canyon
> —a few grains at a time.

As the primary lesson of geology, this pamphlet proclaims
that big results arise as the accumulated effect of tiny
changes. My hand rubbing the canyon wall is an adequate
(if anything, overeffective) illustration of rates that carved
the canyon itself. Time, geology's inexhaustible resource,

performs all the miracles.

Yet, when the pamphlet turns to details, we encounter a different scenario for erosion in Arches. We learn that a balanced rock known as "Chip Off the Old Block" fell during the winter of 1975–76. Before and after photographs of the magnificent Skyline Arch receive the following commentary: "It remained thus for as long as man knew the arch, until, late in 1940, the block of stone fell, and Skyline was suddenly twice its former size." The arches form by sudden, intermittent collapse and toppling, not by imperceptible removal of sand grains. Yet gradualist orthodoxy is so entrenched that the authors of this pamphlet failed to note the inconsistency between their own factual account and the stated theory of their introduction. In other essays of this section, I argue that gradualism is a culturally conditioned prejudice, not a fact of nature, and I make a plea for pluralism in concepts of rate. Punctuational change is at

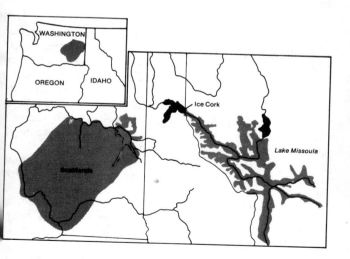

The channeled scablands of eastern Washington.

least as important as imperceptible accumulation. In this essay I tell a local, geologic story. But it conveys the same message—that dogmas play their worst role when they lead scientists to reject beforehand a counterclaim that could be tested in nature.

Flow basalts of volcanic origin blanket most of eastern Washington. These basalts are often covered by a thick layer of loess, a fine-grained, loosely packed sediment blown in by winds during the ice ages. In the area between Spokane and the Snake and Columbia rivers to the south and west, many spectacular, elongate, subparallel channelways are gouged through the loess and deeply into the hard basalt itself. These coulees, to use the local name, must have been conduits for glacial meltwaters, for they run down gradient from an area near the southern extent of the last glacier into the two major rivers of eastern Washington. The channeled scablands—as geologists designate the entire area—are puzzling as well as awesome, and for several reasons:

1. The channels connect across tall divides that once separated them. Since the channels are hundreds of feet deep, this extensive anastomosis indicates that a prodigious amount of water must once have flowed over the divide.
2. As another item favoring channels filled to the brim with water, the sides of the coulees contain many hanging valleys where tributaries enter the main channels. (A hanging valley is a tributary channel that enters a main channel high above the main channel's modern stream bed.)
3. The hard basalt of the coulees is deeply gouged and scoured. This pattern of erosion does not look like the work of gentle rivers in the gradualist mode.
4. The coulees often contain a number of high-standing hills composed of loess that has not been stripped away. These are arranged as if they were once islands in a gigantic braided stream.
5. The coulees contain discontinuous deposits of basaltic stream gravel, often composed of rock foreign to the local area.

Just after World War I, Chicago geologist J Harlen Bretz advanced an unorthodox hypothesis to account for this unusual topography (yes, that's J without a period, and don't ever let one slip in, for his wrath can be terrible). He argued that the channeled scablands had been formed all at once by a single, gigantic flood of glacial meltwater. This local catastrophe filled the coulees, cut through hundreds of feet of loess and basalt, and then receded in a matter of days. He ended his major work of 1923 with these words:

> Fully 3,000 square miles of the Columbia Plateau were swept by the glacial flood, and the loess and silt cover removed. More than 2,000 square miles of this area were left as bare, eroded rock-cut channel floors, now the scablands, and nearly 1,000 square miles carry gravel deposits derived from the eroded basalt. It *was* a debacle which swept the Columbia Plateau.

Bretz's hypothesis became a minor *cause célèbre* within geological circles. Bretz's stout and lonely defense of his catastrophic hypothesis won some grudging admiration, but virtually no support at first. The "establishment," as represented by the United States Geological Survey, closed ranks in opposition. They had nothing better to propose, and they did admit the peculiar character of scabland topography. But they held firm to the dogma that catastrophic causes must never be invoked so long as any gradualist alternative existed. Instead of testing Bretz's flood on its own merits, they rejected it on general principles.

On January 12, 1927, Bretz bearded the lion in its lair and presented his views at the Cosmos Club, in Washington, D.C., before an assembled group of scientists, many from the Geological Survey. The published discussion clearly indicates that a priori gradualism formed the basis for Bretz's glacial reception. I include typical comments from all detractors.

W. C. Alden admitted "it is not easy for one, like myself, who has never examined this plateau to supply offhand an alternative explanation of the phenomena." Nonetheless, undaunted, he continued: "The main difficulties seem to be: (1) The idea that all the channels must have been developed simultaneously in a very short time; and (2) the tremendous amount of water that he postulates. . . . The problem would be easier if less water was required and if longer time and repeated floods could be allotted to do the work."

James Gilluly, this century's chief apostle of geological gradualism, ended a long comment by noting "that the actual floods involved at any given time were of the order of magnitude of the present Columbia's or at most a few times as large, seems by no means excluded by any evidence as yet presented."

E. T. McKnight offered a gradualist alternative for the gravels: "This writer believes them to be the normal channel deposits of the Columbia during its eastward shift over the area in preglacial, glacial, and postglacial times."

G. R. Mansfield doubted that "so much work could be done on basalt in so short a time." He also proposed a calmer explanation: "The scablands seem to me better explained as the effects of persistent ponding and overflow of marginal glacial waters, which changed their position or their places of outlet from time to time through a somewhat protracted period."

Finally, O. E. Meinzer admitted that "the erosion features of the region are so large and bizarre that they defy description." They did not, however, defy gradualist explanation. "I believe the existing features can be explained by assuming normal stream work of the ancient Columbia River." Then, more baldly than most of his colleagues, he proclaimed his faith: "Before a theory that requires a seemingly impossible quantity of water is fully accepted, every effort should be made to account for the existing features without employing so violent an assumption."

The story has a happy ending, at least from my point of view, for Bretz was delivered from the lion's lair by later evidence. Bretz's hypothesis has prevailed, and virtually all geologists now believe that catastrophic floods cut the channeled scablands. Bretz had found no adequate source for his floodwaters. He knew that the glaciers had advanced as far as Spokane, but neither he nor anyone else could imagine a reasonable way to melt so much water so rapidly. Indeed, we still have no mechanism for such an episodic melting.

The solution came from another direction. Geologists found evidence for an enormous, ice-dammed glacial lake in western Montana. This lake emptied catastrophically when the glacier retreated and the dam broke. The spillway for its waters leads right into the channeled scablands.

Bretz had presented no really direct evidence for deep, surging water. Gouging might have proceeded sequentially, rather than all at once; anastomosis and hanging valleys might reflect filled coulees with gentle, rather than raging, flow. But when the first good aerial photographs of the scablands were taken, geologists noticed that several areas on the coulee floors are covered with giant stream bed ripples, up to 22 feet high and 425 feet long. Bretz, like an ant on a Yale bladderball, had been working on the wrong scale. He had been walking over the ripples for decades but had been too close to see them. They are, he wrote quite correctly, "difficult to identify at ground level under a cover of sagebrush." Observations can only be made at appropriate scales.

Hydraulic engineers can infer the character of flow from the size and shape of ripples on a stream bed. V. R. Baker estimates a maximum discharge of 752,000 cubic feet per second in the scabland flow channels. Such a flood could have moved 36-foot boulders.

I could end here with a cardboard version of the story much to my liking: Perceptive hero suppressed by blinded dogmatists stands firm, expresses his allegiance to fact over received opinion, and eventually prevails by patient persuasion and overwhelming documentation. The outline of this tale is surely valid: gradualist bias *did* lead to a rejection of Bretz's catastrophic hypothesis out of hand, and Bretz (apparently) was right. But, as I read through the original papers, I realized that this good guy–bad guy scenario must yield to a more complex version. Bretz's opponents were not benighted dogmatists. They did have a priori preferences, but they also had good reasons to doubt catastrophic flooding based on Bretz's original arguments. Moreover, Bretz's style of scientific inquiry virtually guaranteed that he would not triumph with his initial data.

Bretz proceeded in the classic tradition of strict empiricism. He felt that adventurous hypotheses could only be established by long and patient collecting of information in the field. He eschewed theoretical discussion and worried little about the valid conceptual problem that so bothered his adversaries: where could so much water come from so suddenly?

Bretz tried to establish his hypothesis by toting up evidence of erosion in the field, piece by patient piece. He seemed singularly uninterested in finding the missing item that would render his story coherent—a source for the water. For this attempt might involve speculation without direct evidence, and Bretz relied only upon fact. When Gilluly challenged him on the absence of a source for the water, Bretz simply replied: "I believe that my interpretation of channeled scabland should stand or fall on the scabland phenomena themselves."

But why should an opponent be converted by such an incomplete theory? Bretz believed that the southern end of the glacier had melted precipitously, but no scientist could imagine a way to melt ice so quickly. (Bretz tentatively suggested volcanic activity under the ice, but quickly abandoned the theory when Gilluly attacked.) Bretz stayed in the scablands, while the answer sat in western Montana. Glacial Lake Missoula had been in the literature since the 1880s, but Bretz did not make the connection—he was working in other ways. His opponents were right. We still do not know a way to melt so much ice so quickly. But the premise shared by all participants was wrong: the source of the water was water.

Events that "cannot happen" according to received wisdom rarely gain respectability by a simple accumulation of evidence for their occurrence; they require a mechanism to explain how they *can* happen. Early supporters of continental drift ran into the same difficulty that Bretz encountered. Their evidence of faunal and lithological similarities between continents now widely separated strikes us today as overwhelming, but it failed in their time because no reasonable force had been proposed for moving continents. The theory of plate tectonics has since provided a mechanism and established the idea of continental drift.

Moreover, Bretz's opponents did not rest their case entirely on the unorthodox character of Bretz's hypothesis. They also marshaled some specific facts on their side, and they were partly right. Bretz originally insisted upon a single flood, while his opponents cited much evidence to show that the scablands had not formed all at once. We now know that Lake Missoula formed and re-formed several times as the glacial margin fluctuated. In his latest work, Bretz called for eight separate episodes of catastrophic flooding. Bretz's opponents were wrong in inferring gradual change from the evidence of temporal spread: catastrophic episodes can be separated by long periods of quiescence. But Bretz was also wrong in attributing the formation of the scablands to a single flood.

I prefer heroes of flesh, blood, and fallibility, not of tinseled cardboard. Bretz is inscribed on my ledger because he stood against a firm, highly restrictive dogma that never had made any sense: the emperor had been naked for a century. Charles Lyell, the godfather of geological gradualism, had pulled a fast one in establishing the doctrine of imperceptible change. He had argued, quite rightly, that geologists must invoke the invariance (uniformity) of natural law through time in order to study the past scientifically. He then applied the same term—uniformity—to an empirical claim about rates of processes, arguing that change must be slow, steady, and gradual, and that big results can only arise as the accumulation of small changes.

But the uniformity of law does not preclude *natural* catastrophes, particularly on a local scale. Perhaps some invariant laws operate to produce infrequent episodes of sudden, profound change. Bretz may not have cared for this brand of philosophical waffling. He probably would brand it as vacuous nonsense preached by an urban desk man. But he had the independence and gumption to live by a grand old slogan from Horace, often espoused by science but not often followed: *Nullius addictus jurare in verba magistri,* "I am not bound to swear allegiance to the words of any master."

My tale ends with two happy postscripts. First, Bretz's hypothesis that channeled scabland reflects the action of catastrophic flooding has been fruitful far beyond Bretz's local area. Scablands have been found in association with other western lakes, most notably Lake Bonneville, the large ancestor of a little puddle in comparison—Great Salt Lake, Utah. Other applications have ranged about as far as they can go. Bretz has become the darling of planetary geologists who find in the channelways of Mars a set of features best interpreted by Bretz's style of catastrophic flooding.

Second, Bretz did not share the fate of Alfred Wegener, dead on the Greenland ice while his theory of continental drift lay in limbo. J Harlen Bretz presented his hypothesis sixty years ago, but he has lived to enjoy his vindication. He is now well into his nineties, feisty as ever and justly pleased with himself. In 1969, he published a forty-page paper summarizing a half century of controversy about the channeled scablands of eastern Washington. He closed with this statement:

The International Association for Quaternary Research held its 1965 meeting in the United States. Among the many field excursions it organized was one in the northern Rockies and the Columbia Plateau in Washington. . . . The party . . . traversed the full length of the Grand Coulee, part of the Quincy basin and much of the Palouse-Snake scabland divide, and the great flood gravel deposits in the Snake Canyon. The writer, unable to attend, received the next day a telegram of "greetings and salutations" which closed with the sentence, "We are now all catastrophists."

Postscript

I sent a copy of this article to Bretz after its publication in *Natural History.* He replied on October 14, 1978:

Dear Mr. Gould,

Your recent letter is most gratifying. Thank you for understanding.

I have been surprised by the way my pioneer Scabland work has been applauded and further developed. I knew all along that I was right but the decades of doubt and challenge had produced an emotional lethargy, I think. Then the surprise following Victor Baker's field trip in June woke me up again. What! Had I become a semi-authority on extra-terrestrial processes and events?

Physically incapacitated now (I am 96), I can only cheer the work of others in a field where I was a pathfinder. Again I thank you.

J Harlen Bretz

In November 1979, at the annual meeting of the Geological Society of America, the Penrose Medal (the profession's premier award) was given to J Harlen Bretz.

20 | A Quahog Is a Quahog

THOMAS HENRY HUXLEY once defined science as "organized common sense." Other contemporaries, including the great geologist Charles Lyell, urged an opposing view—science, they said, must probe behind appearance, often to combat the "obvious" interpretation of phenomena.

I cannot offer any general rules for the resolution of conflicts between common sense and the dictates of a favored theory. Each camp has won its battles and received its lumps. But I do want to tell a story of common sense triumphant—an interesting story because the theory that seemed to oppose ordinary observation is also correct, for it is the theory of evolution itself. The error that brought evolution into conflict with common sense lies in a false implication commonly drawn from evolutionary theory, not with the theory itself.

Common sense dictates that the world of familiar, macroscopic organisms presents itself to us in "packages" called species. All bird watchers and butterfly netters know that they can divide the specimens of any local area into discrete units blessed with those Latin binomials that befuddle the uninitiated. Occasionally, to be sure, a package may become unraveled and even seem to coalesce with another. But such cases are noted for their rarity. The birds of Massachusetts and the bugs in my backyard are unambiguous members of species recognized in the same way by all experienced observers.

This notion of species as "natural kinds" fit splendidly with creationist tenets of a pre-Darwinian age. Louis Agassiz even argued that species are God's individual thoughts, made incarnate so that we might perceive both His majesty and His message. Species, Agassiz wrote, are "instituted by the Divine Intelligence as the categories of his mode of thinking."

But how could a division of the organic world into discrete entities be justified by an evolutionary theory that proclaimed ceaseless change as the fundamental fact of nature? Both Darwin and Lamarck struggled with this question and did not resolve it to their satisfaction. Both denied to the species any status as a natural kind.

Darwin lamented: "We shall have to treat species as . . . merely artificial combinations made for convenience. This may not be a cheering prospect; but we shall at least be freed from the vain search for the undiscovered and undiscoverable essence of the term species." Lamarck complained: "In vain do naturalists consume their time in describing new species, in seizing upon every nuance and slight peculiarity to enlarge the immense list of described species."

Yet—and this is the irony—both Darwin and Lamarck were respected systematists who named hundreds of species. Darwin wrote a four-volume taxonomic treatise on barnacles, while Lamarck produced more than three times as many volumes on fossil invertebrates. Faced with the practicum of their daily work, both recognized entities where theory denied their reality.

There is a traditional escape from this dilemma: one can argue that our world of ceaseless flux alters so slowly that configurations of the moment may be treated as static. The coherence of modern species disappears through time as they transform slowly into their descendants. One can only remember Job's lament about "man that is born of a woman"—"He cometh forth like a flower . . . he fleeth also as a shadow, and continueth not." But Lamarck and Darwin could not even enjoy this resolution, for they both worked extensively with fossils and were as successful in dividing evolving sequences into species as they were in parsing the modern world.

Other biologists have even forsworn this traditional escape and denied the reality of species in any context. J.B.S. Haldane, perhaps the most brilliant evolutionist of this century, wrote: "The concept of a species is a concession to our linguistic habits and neurological mechanisms." A paleontological colleague proclaimed in 1949 that "a species . . . is a fiction, a mental construct without objective existence."

Yet common sense continues to proclaim that, with few exceptions, species can be clearly identified in local areas of our modern world. Most biologists, although they may deny the reality of species through geologic time, do affirm their status for the modern moment. As Ernst Mayr, our leading student of species and speciation, writes: "Species are the product of evolution and not of the human mind." Mayr argues that species are "real" units in nature as a result both of their history and the current interaction among their members.

Species branch off from ancestral stocks, usually as small, discrete populations inhabiting a definite geographical area. They establish their uniqueness by evolving a genetic program sufficiently distinct that members of the species breed with each other, but not with members of other species. Their members share a common ecological niche and continue to interact through interbreeding.

Higher units of the Linnaean hierarchy cannot be objectively defined, for they are collections of species and have no separate existence in nature—they neither interbreed nor necessarily interact at all. These higher units—genera, families, orders, and on up—are not arbitrary. They must not be inconsistent with evolutionary genealogy (you cannot put people and dolphins in one order and chimps in another). But ranking is, in part, a matter of custom with no "correct" solution. Chimps are our closest relatives by genealogy, but do we belong in the same genus or in different genera within the same family? Species are nature's only objective taxonomic units.

Shall we then follow Mayr or Haldane? I am a partisan of Mayr's view and I wish to defend it with an offbeat but, to my mind, persuasive line of evidence. The repeated experiment is a cornerstone of scientific methods—although evolutionists, dealing with nature's uniqueness, do not often have an opportunity to practice it. But in this case, we have a way to obtain valuable information about whether species are mental abstractions embedded in cultural practice or packages in nature. We can study how different peoples, in complete independence, divide the organisms of their local areas into units. We can contrast Western classifications into Linnaean species with the "folk taxonomies" of non-Western peoples.

The literature on non-Western taxonomies is not extensive, but it is persuasive. We usually find a remarkable correspondence between Linnaean species and non-Western plant and animal names. In short, the same packages are recognized by independent cultures. I do not argue that folk taxonomies invariably include the entire Linnaean catalog. People usually do not classify exhaustively unless organisms are important or conspicuous. The Fore of New Guinea have a single word for all butterflies, although species are as distinct as the birds they do classify in Linnaean detail. Similarly, most of the bugs in my backyard have no common name in our folk taxonomy, but all the birds in Massachusetts do. The Linnaean correspondences only arise when folk taxonomies attempt an exhaustive division.

Several biologists have noted these remarkable correspondences in the course of their fieldwork. Ernst Mayr himself describes his experience in New Guinea: "Forty years ago, I lived all alone with a tribe of Papuans in the mountains of New Guinea. These superb woodsmen had 136 names for the 137 species of birds I distinguished (confusing only two nondescript species of warblers). That . . . Stone Age man recognizes the same entitites of nature as Western university-trained scientists refutes rather decisively the claim that species are nothing but a product of the human imagination." In 1966, Jared Diamond published a more extensive study on the Fore people of New Guinea. They have names for all the Linnaean bird species in their area. Moreover, when Diamond brought seven Fore men into a new area populated by birds they had never seen, and asked them to give the closest Fore equivalent for each new bird, they placed 91 of 103 species into the Fore group closest to the new species in our Western Linnaean classification. Diamond relates an interesting tale:

> One of my Fore assistants collected a huge, black, short-winged, ground-dwelling bird, which neither he nor I had seen before. While I was puzzled by its affinities, the Fore man promptly proclaimed it to be a *peteobeye,* the name for a graceful little brown cuckoo which frequents trees in Fore gardens. The new bird eventually proved to be Menbek's coucal, an aberrant member of the cuckoo family, to which some features of body form and leg and bill shape betray its affinity.

These informal studies by biologists have been supplemented in recent years with two exhaustive treatments by anthropologists who are also competent natural historians —Ralph Bulmer's work on vertebrate taxonomies of the Kalam people of New Guinea, and Brent Berlin's study (with botanists Dennis Breedlove and Peter Raven) of plant classification by the Tzeltal Indians of highland Chiapas, Mexico. (I thank Ernst Mayr for introducing me to Bulmer's work and for urging this line of argument for many years.)

The Kalam people, for example, use frogs extensively as food. Most of their frog names have a one-to-one correspondence with Linnaean species. In some cases they apply the same name to more than one species, but still recognize the difference: Kalam informants could readily identify two different kinds of *gunm,* distinguished both by appearance and habitat, even though they had no standard names for them. Sometimes, the Kalam do better than we. They recognize, as *kasoj* and *wyt,* two species that had been lumped incorrectly under the single Western name *Hyla becki.*

Bulmer has recently teamed up with Ian Saem Majnep, a Kalam, to produce a remarkable book, *Birds of My Kalam Country.* More than 70 percent of Saem's names have one-to-one correspondence with Western species. In most other cases, he either lumps two or more Linnaean species under the same Kalam name but recognizes the Western distinction, or else he makes divisions within a Western species but recognizes the unity (in some birds of paradise, for example, he names the sexes separately because only males carry the prized plumage). In only one case does Saem follow a practice inconsistent with Linnaean nomenclature—he uses the same name for drab females in two species of birds of paradise, but awards different names to the showy males of each species. In fact, Bulmer could only find four cases (2 percent) of inconsistency in the entire Kalam catalog of 174 vertebrate species, spanning mammals, birds, reptiles, frogs, and fishes.

Berlin, Breedlove, and Raven published their first study in 1966, explicitly to challenge Diamond's claim for the generality of extensive one-to-one correspondence between folk names and Linnaean species. They held initially that only 34 percent of Tzeltal plant names matched Linnaean species and that a large variety of "misclassifications" reflected cultural uses and practices. But a few years later, in a frank article, they reversed their opinion and affirmed the uncannily close correspondence of folk and Linnaean names. They had, in the earlier study, not fully understood the Tzeltal system of hierarchical ordering and had mixed names from several levels in establishing the basic folk groups. In addition, Berlin admitted that he had been led astray by a standard anthropological bias for cultural relativism. I cite his recantation, not to show him up, but as a token of my admiration for an act all too rarely performed by scientists (although any scientist worth his salt has changed his mind about fundamental issues):

Many anthropologists, whose traditional bias is to see the total relativity of man's variant classifications of reality, have generally been hesitant to accept such findings. . . . My colleagues and I, in an earlier paper, have presented arguments in favor of the "relativist" view. Since the publication of that report more data have been made available, and it now appears that this position must be seriously reconsidered. There is at present a growing body of evidence that suggests that the fundamental taxa recognized in folk systematics correspond fairly closely with scientifically known species.

Berlin, Breedlove, and Raven have now published an exhaustive book on Tzeltal taxonomy, *Principles of Tzeltal Plant Classification*. Their complete catalog contains 471 Tzeltal names. Of these, 281, or 61 percent, stand in one-to-one correspondence with Linnaean names. All but 17 of the rest are, in the authors' terms, "underdifferentiated"—that is, the Tzeltal names refer to more than one Linnaean species. But, in more than two-thirds of these cases, the Tzeltal use a subsidiary system of naming to make distinctions within the primary groups, and all these subsidiaries correspond with Linnaean species. Only 17 names, or 3.6 percent, are "overdifferentiated" by referring to part of a Linnaean species. Seven Linaean species have two Tzeltal names, and only one has three—the bottle gourd *Lagenaria siceraria*. The Tzeltal distinguish bottle gourd plants by the utility of their fruits—one name for large, round fruits used as containers for tortillas; another for long-necked gourds well suited for carrying liquids; and a third for small, oval fruits that are not used at all.

A second, equally interesting generality emerges from studies of folk classification. Biologists argue that only species are real units in nature, and that names at higher levels of the taxonomic hierarchy represent human decisions about how species should be grouped (under the constraint, of course, that such grouping be consistent with evolutionary genealogy). Thus, for names applied to groups of species, we should not expect one-to-one correspondence with Linnaean designations but should anticipate a variety of schemes based upon local uses and culture. Such variety has been a consistent finding in studies of folk taxonomy. Groups of species often include basic forms attained independently by several evolutionary lines. The Tzeltal, for example, have four broader names for groups of species, roughly corresponding to trees, vines, grasses, and broad-leafed herbaceous plants. These names apply to about 75 percent of their plant species, while others, like corn, bamboo, and agave, are "unaffiliated."

Often, the grouping of species reflects more subtle and pervasive aspects of culture. The Kalam of New Guinea, for example, divide their nonreptilian four-footed vertebrates into three classes: *kopyak,* or rats; *kmn* for an evolutionarily heterogeneous collection of larger game mammals, mostly marsupials and rodents; and *as* for an even more heterogeneous collection of frogs and small rodents. (Under repeated questioning by Bulmer, the Kalam deny any subdivision between frogs and rodents within *as,* although they do acknowledge [and dismiss as unimportant] the morphological similarity between small furry *as* and rodents among *kmn.* They also recognize that some *kmn* have pouches and others do not.) The divisions reflect fundamental facts of Kalam culture. *Kopyak,* associated with excrement and unclean food around homesteads, are not eaten at all. *As* are collected primarily by women and children and, although eaten by most men and collected by some, are forbidden foods for boys during their rites of passage and for adult men who practice sorcery. *Kmn* are hunted primarily by men.

Likewise, birds and bats are all *yakt,* with the single exception of the large, flightless cassowary called *kobty.* The distinction is made for deeper and more complex reasons than mere appearance—for the Kalam do recognize avian characters in *kobty.* Cassowaries, Bulmer argues, are the prime game of the forest and the Kalam maintain an elaborate cultural antithesis between cultivation (represented by taro and pigs) and the forest (represented by pandanus nuts and cassowaries). Cassowaries are also the mythological sisters of men.

We maintain similar practices in our own folk taxonomy. Edible mollusks are "shellfish," but Linnaean species all have common names. I well remember the reprimand I received from a New England shipmate when I applied the informal scientific term "clam" to all bivalved mollusks (to him a clam is only the steamer, *Mya arenaria*): "A quahog is a quahog, a clam is a clam, and a scallop is a scallop."

The evidence of folk taxonomy is persuasive for the modern world. Unless the tendency to divide organisms into Linnaean species reflects a neurological style wired into all of us (an interesting proposition, but one that I doubt), the world of nature is, in some fundamental sense, really divided into reasonably discrete packages of creatures as a result of evolution. (I do not, of course, deny that our propensity for classifying in the first place reflects something about our brains, their inherited capacities, and the limited ways in which complexity may be ordered and made sensible. I merely doubt that such a definite procedure as classification into Linnaean species could reflect the constraints of our mind alone, and not of nature.)

But are these Linnaean species, recognized by independent cultures, merely temporary configurations of the moment, mere way stations on evolutionary lineages in continual flux? I argue in essays 17 and 18 that, contrary to popular belief, evolution does not work this way, and that species have a "reality" through time to match their distinctness at a moment. An average species of fossil invertebrates lives five to ten million years (terrestrial vertebrates have shorter average durations). During this time, they rarely change in any fundamental way. They become extinct, without issue, looking much as they did when they first appeared.

New species usually arise, not by the slow and steady transformation of entire ancestral populations, but by the splitting off of small isolates from an unaltered parental stock. The frequency and speed of such speciation is among the hottest topics in evolutionary theory today, but I think that most of my colleagues would advocate ranges of hundreds of thousands of years for the origin of most species by splitting. This may seem like a long time in the framework of our lives, but it is a geological instant, usually represented in the fossil record by a single bedding plane, not a long stratigraphic sequence. If species arise in hundreds or thousands of years and then persist, largely unchanged, for several million, the period of their origin is a tiny fraction of one percent of their total duration. Therefore, they may be treated as discrete entities even through time. Evolution at higher levels is fundamentally a story of the differential success of species not the slow transformation of lineages.

Of course, if we happen to encounter a species during the geological microsecond of its origin, we will not be able to make clear distinctions. But our chances of finding a species in this state are small indeed. Species are stable entities with very brief periods of fuzziness at their origin (although not at their demise because most species disappear cleanly without changing into anything else). As Edmund Burke said in another context: "Though no man can draw a stroke between the confines of day and night, yet light and darkness are upon the whole tolerably distinguishable."

Evolution is a theory of organic change, but it does not imply, as many people assume, that ceaseless flux is the irreducible state of nature and that structure is but a temporary incarnation of the moment. Change is more often a rapid transition between stable states than a continuous transformation at slow and steady rates. We live in a world of structure and legitimate distinction. Species are the units of nature's morphology.

6 | Early Life

21 | An Early Start

P O O H - B A H , T H E Lord High Everything Else of Titipu, boasted a family pride so strong as to be "something inconceivable." "You will understand this," he said to Nanki-Poo in suggesting that a bribe would be both appropriate and expensive, "when I tell you that I can trace my ancestry back to a protoplasmal primordial atomic globule."

If human pride is nurtured by such vastly extended roots, then the end of 1977 was a bounteous time for self-esteem. Early in November, an announcement of the discovery of some fossil prokaryotes from South Africa pushed the antiquity of life back to 3.4 billion years. (Prokaryotes, including bacteria and blue green algae, form the kingdom Monera. Their cells contain no organelles—no nucleus, no mitochondria—and they are regarded as the simplest forms of life on earth.) Two weeks later, a research team from the University of Illinois announced that the so-called methane-producing bacteria are not closely related to other monerans after all, but form a separate kingdom of their own.

If true monerans were alive 3.4 billion years ago, then the common ancestor of monerans and these newly christened "methanogens" must be considerably more ancient. Since the oldest dated rocks, the Isua Supracrustals of West Greenland, are 3.8 billion years old, we are left with very little time between the development of suitable conditions for life on the earth's surface and the origin of life itself. Life is not a complex accident that required immense time to convert the vastly improbable into the nearly certain—to build laboriously, step by step, through a large chunk of time's vastness, the most elaborate machinery on earth from the simple constituents of our original atmosphere. Instead, life, for all its intricacy, probably arose rapidly about as soon as it could; perhaps it was as inevitable as quartz or feldspar. (The earth is some 4½ billion years old, but it passed through a molten or near-molten stage some time after its formation and probably did not form a solid crust much before the deposition of the West Greenland sequence.) No wonder these stories hit the front page of the *New York Times,* and even inspired an editorial for Veterans' Day musings.

Twenty years ago, I spent a summer at the University of Colorado, fortifying myself for the transition from high school to college. Amidst the various joys of snowcapped peaks and sore asses from trying to "set a trot," I well remember the highlight of my stay—George Wald's lecture on the "Origin of Life." He presented with infectious charm and enthusiasm the perspective that developed in the early 1950s and reigned as an orthodoxy until very recently.

In Wald's view, the spontaneous origin of life could be considered as a virtually inevitable consequence of the earth's atmosphere and crust, and of its favorable size and position in the solar system. Still, he argued, life is so staggeringly complex that its origin from simple chemicals must have consumed an immense amount of time—probably more time than its entire subsequent evolution from DNA molecule to advanced beetles (or whatever you choose to place atop the subjective ladder). Thousands of steps, each requiring the one before, each improbable in itself. Only the immensity of time guaranteed the result, for time converts the improbable to the inevitable—give me a million years and I'll flip a hundred heads in a row more than once Wald wrote in 1954: "Time is in fact the hero of the plot The time with which we have to deal is the order of two billion years. . . . Given so much time, the 'impossible' becomes possible, the possible probable, and the probable virtually certain. One has only to wait: time itself performs the miracles."

This orthodox view congealed without the benefit of any direct data from paleontology to test it, for the paucity of fossils before the great Cambrian "explosion" 600 million years ago is, perhaps, the outstanding fact and frustration of my profession. In fact, the first unambiguous evidence of Precambrian life appeared in the same year that Wald theorized about its origin. Harvard paleobotanist Elso Barghoorn and Wisconsin geologist S. A. Tyler described a series of prokaryotic organisms from cherts of the Gunflint Formation, rocks nearly two billion years old from the northern shore of Lake Superior. Still, the gap between the Gunflint and the earth's origin spanned 2½ billion years, more than enough time for Wald's slow and steady construction.

But our knowledge of life continued its trek backward. Laminated carbonate deposits, called stromatolites, had been known for some time from rocks of the Bulawayan Series, 2.6 to 2.8 billion years old, in Southern Rhodesia. The laminations resemble patterns formed by modern blue green algal mats that trap and bind sediment. The organic interpretation of stromatolites won many converts after Barghoorn and Tyler's Gunflint discoveries removed the odor of heresy from belief in Precambrian fossils. Then, ten years ago in 1967, Barghoorn and J. W. Schopf reported "algalike" and "bacteriumlike" organisms from the Fig Tree Series of South Africa. Now the orthodox idea of slow construction spanning most of the earth's history began to rumble, for the Fig Tree rocks, based on dates available in 1967, seemed to be more than 3.1 billion years old. Schopf and Barghoorn dignified their discoveries with formal Latin names, but their own characterizations—algalike and bacteriumlike—reflected their doubts. In fact, Schopf later decided that the balance of evidence stood against the biological nature of these structures.

The recent announcement of 3.4-billion-year-old life is not a startlingly new discovery, but a satisfactory culmination of a decade's debate about the status of life in the Fig

Tree. The new evidence, gathered by Andrew H. Knoll and Barghoorn, also comes from cherts of the Fig Tree Series. But now the evidence is close to conclusive; moreover, recent dates indicate a greater age of 3.4 billion years for the series. In fact, the Fig Tree cherts may be the oldest appropriate rocks on earth for the discovery of ancient life. Older Greenland rocks have been too altered by heat and pressure to preserve organic remains. Knoll tells me that some unstudied cherts in Rhodesia may range back to 3.6 billion years, but eager scientists will have to await a political denouement before their arcane concerns attract sympathy or ensure safety. Still, the notion that life has been found in the oldest rocks that could contain evidence of it forces us, I think, to abandon the view of life's slow, steady, and improbable development. Life arose rapidly, perhaps as soon as the earth cooled down sufficiently to support it.

The new fossils from the Fig Tree Series are far more convincing than the previous discoveries. "In younger rocks [they] would without hesitation be called algal microfossils," Knoll and Barghoorn claim. This interpretation rests upon five arguments:

1. The new structures are within the size range of modern prokaryotes. The earlier structures described by Schopf and Barghoorn were disturbingly large; Schopf later rejected them as biological, primarily on the basis of their large size. The new fossils, averaging 2.5 micrometers in diameter (a micrometer is a millionth of a meter), have a mean volume only 0.2 percent as large as the earlier structures now considered inorganic.

2. Populations of modern prokaryotes have a characteristic distribution of size. They can be arranged in a typical bell-shaped curve, with the average diameter most frequent and a continual decrease in number towards larger or smaller sizes. Thus, prokaryotic populations not only have a diagnostic average size (point 1 above), they also have a characteristic pattern of variation about this average. The new microfossils form a beautiful bell-shaped distribution with limited spread (range from 1 to 4 micrometers). The previous, larger structures exhibited much greater variation and no strong mean.

3. The new structures are "variously elongated, flattened, wrinkled, or folded" in a manner strikingly similar to Gunflint and later Precambrian prokaryotes. Such shapes are characteristic of postmortem degradation in modern prokaryotes. The larger, earlier structures were distressingly spherical; spheres, as a standard configuration of minimal surface area, can be easily produced by a host of inorganic processes—consider bubbles.

4. Most convincingly, about one quarter of the new microfossils have been found in various stages of cell division. Lest such a proportion caught *in flagrante delicto* sems unreasonably high, I point out that prokaryotes can divide every twenty minutes or so and take several minutes to complete the process. A single cell might well spend one-fourth of its life making two daughters.

5. These four arguments based on morphology are persuasive enough for me, but Knoll and Barghoorn add some biochemical evidence as well. Atoms of a single element often exist in several alternate forms of different weight. These forms, called isotopes, have the same number of protons but different numbers of neutrons. Some isotopes are radioactive and break down spontaneously to other elements; others are stable and persist unchanged throughout geologic time. Carbon has two major stable isotopes, C^{12} with 6 protons and 6 neutrons, and C^{13} with 6 protons and 7 neutrons. When organisms fix carbon in photosynthesis, they use preferentially the lighter isotope C^{12}. Hence, the C^{12}/C^{13} ratio of carbon fixed by photosynthesis is higher than the ratio in inorganic carbon (in a diamond, for example). Moreover, since both isotopes are stable, their ratio will not alter through time. The C^{12}/C^{13} ratios for Fig Tree carbon are too high for an inorganic origin; they are in the range for fixation by photosynthesis. This, in itself, would not establish the case for life in the Fig Tree; light carbon can be fixed preferentially in other ways. But combined with the evidence of size, distribution, shape, and cellular division, this additional support from biochemistry completes a convincing case.

If prokaryotes were well established 3.4 billion years ago, now much further back shall we seek the origin of life? I have already pointed out that no suitable (or at least accessible) older rocks are known on earth, so for now we can proceed no further from the direct evidence of fossils. We turn instead to the second front-page item, the claim of Carl Woese and his associates that methanogens are not bacteria at all, but may represent a new kingdom of prokaryotic life, distinct from the Monera (bacteria and blue green algae). Their report has been widely distorted, most notably in the *New York Times* editorial of November 11, 1977. The *Times* proclaimed that the great dichotomy of plants and animals had finally been broken: "Every child learns about things being vegetable or animal—a division as universal as the partition of mammals into male and female. Yet . . . [we now have] a 'third kingdom' of life on earth, organisms that are neither animal nor vegetable, but of another category altogether." But biologists abandoned "the great dichotomy" long ago, and no one now tries to cram all single-celled creatures into the two great groups traditionally recognized for complex life. Most popular these days is a system of five kingdoms: plants, animals, fungi, protists (single-celled eukaryotes, including amoebas and paramecia, with nucleus, mitochondria, and other organelles), and the prokaryotic monerans. If methanogens are promoted, they will form a sixth kingdom, joining the monerans in a superkingdom, Prokaryota. Most biologists regard the division between prokaryotes and eukaryotes, not between plants and animals, as the fundamental partition of life.

Woese's research group (see Fox, *et al.*, 1977 in the bibliography) isolated a common RNA from ten methanogens and from three monerans for comparison (DNA makes RNA, and RNA serves as the template upon which proteins are synthesized). A single strand of RNA, like DNA, consists of a sequence of nucleotides. Any one of four nucleotides can occupy each position, and each group of three nucleotides specifies an amino acid; proteins are built of amino acids arranged in folded chains. This, in a compressed phrase, is the "genetic code." Biochemists can now "sequence" RNA, that is, they can read the entire sequence of nucleotides in order down the RNA strand.

The prokaryotes (methanogens, bacteria, and blue-green algae) must have had a common ancestor at some time near the origin of life. Thus, all prokaryotes had the same RNA sequence at one point in their past; any current differences arose by divergence from this common ancestral sequence, after the trunk of the prokaryotic tree split up into its several branches. If molecular evolution proceeded at a constant rate, then the extent of current difference between any two forms would directly record the amount of time since their lineages split from a common ancestor—that is, the last time they shared the same RNA sequence. Perhaps, for example, a different nucleotide in the two forms at 10 percent of all common positions would indicate a time of divergence a billion years ago; 20 percent, two billion years, and so on.

Woese and his group measured the RNA differences for all pairs of species among the ten methanogens and three monerans and used the results to construct an evolutionary tree. This tree contains two major limbs—all the methanogens on one, all the monerans on the other. They chose their three monerans to represent the greatest differences within the group—enteric (gut) bacteria versus free-living blue-green algae, for example. Nonetheless, each moneran is more similar to all other monerans than any moneran is to any methanogen.

The simplest interpretation of these results holds that methanogens and monerans are separate evolutionary groups, with a common ancestry preceding the appearance of either. (Previously, methanogens had been classified among the bacteria; in fact, they had not been recognized as a coherent entity at all, but had been regarded as a set of independent evolutionary events—convergent evolution for the ability to make methane). This interpretation underlies Woese's claim that methanogens are separate from monerans and should be recognized as a sixth kingdom. Since good monerans had already evolved by Fig Tree times, 3.4 billion or more years ago, the common ancestry of methanogens and monerans must have been even earlier, thus pushing the origin of life even further back toward the beginning of the earth itself.

This simple interpretation, as Woese and his group realize, is not the only possible reading of their results. We may propose two other perfectly plausible hypotheses: (1) The three monerans that they used may not represent the entire group very well. Perhaps the RNA sequences of other monerans will differ as much from the first three as all the methanogens do. We would then have to include the methanogens with all monerans in a single grand group. (2) The assumption of nearly constant evolutionary rates may not hold. Perhaps the methanogens split off from one branch of monerans long after the main groups of monerans had branched from their common ancestor. These early methanogens may then have evolved at a rate far in excess of that followed by moneran groups in diverging from each other. In this case, the great difference in RNA sequence between any methanogen and any moneran would only record a rapid evolutionary rate for early methanogens, not a common ancestry with monerans before the monerans themselves split into subgroups. The gross amount of biochemical difference will accurately record time of divergence only if evolution proceeds at reasonably constant biochemical rates.

But one other observation makes Woese's hypothesis attractive and inspires my own strong rooting for it. The methanogens are anaerobic; they die in the presence of oxygen. Hence, they are confined today to unusual environments: muds at the bottom of ponds depleted of oxygen or deep hot springs in Yellowstone Park, for example. (The methanogens grow by oxidizing hydrogen and reducing carbon dioxide to methane—hence their name.) Now, amidst all the disagreement that afflicts the study of our early earth and its atmosphere, one point has gained general assent: our original atmosphere was devoid of oxygen and rich in carbon dioxide, the very conditions under which methanogens thrive and for which the earth's original life might have evolved. Could modern methanogens be remnants of the earth's first biota, originally evolved to match its general condition, but now restricted by the spread of oxygen to a few marginal environments? We believe that most free oxygen in our atmosphere is the product of organic photosynthesis. The Fig Tree organisms were already indulging in photosynthesis. Thus, the golden age of methanogens may have passed long before the advent of Fig Tree monerans. If this reverie be confirmed, then life must have originated long before Fig Tree times.

In short, we now have direct evidence of life in the oldest rocks that could contain it. And, by reasonably strong inference, we have reason to believe that a major radiation of methanogens predated these photosynthesizing monerans. Life probably arose about as soon as the earth became cool enough to support it.

Two closing thoughts, admittedly reflecting my personal prejudices: First, as a strong adherent to exobiology, that great subject without a subject matter (only theology may exceed us in this), I am delighted by the thought that life may be more intrinsic to planets of our size, position, and composition than we had ever dared to imagine. I feel even more certain that we are not alone, and I hope that more effort will be directed toward the search for other civilizations by radio-telescope. The difficulties are legion, but a positive result would be the most stupendous discovery in human history.

Secondly, I am led to wonder why the old, discredited orthodoxy of gradual origin ever gained such strong and general assent. Why did it seem so reasonable? Certainly not because any direct evidence supported it.

I am, as several other essays emphasize, an advocate of the position that science is not an objective, truth-directed machine, but a quintessentially human activity, affected by passions, hopes, and cultural biases. Cultural traditions of thought strongly influence scientific theories, often directing lines of speculation, especially (as in this case) when virtually no data exist to constrain either imagination or prejudice. In my own work (see essays 17 and 18), I have been impressed by the powerful and unfortunate influence that gradualism has exerted on paleontology via the old motto *natura non facit saltum* ("nature does not make leaps").

Gradualism, the idea that all change must be smooth, slow, and steady, was never read from the rocks. It represented a common cultural bias, in part a response of nineteenth-century liberalism to a world in revolution. But it continues to color our supposedly objective reading of life's history.

In the light of gradualistic presuppositions, what other interpretation could have been placed upon the origin of life? It is an enormous step from the constituents of our original atmosphere to a DNA molecule. Therefore, the transition must have progressed laboriously through multitudes of intervening steps, one at a time, over billions of years.

But the history of life, as I read it, is a series of stable states, punctuated at rare intervals by major events that occur with great rapidity and help to establish the next stable era. Prokaryotes ruled the earth for three billion years until the Cambrian explosion, when most major designs of multicellular life appeared within ten million years. Some 375 million years later, about half the families of invertebrates became extinct within a few million years. The earth's history may be modelled as a series of occasional pulses, driving recalcitrant systems from one stable state to the next.

Physicists tell us that the elements may have formed during the first few minutes of the big bang; billions of subsequent years have only reshuffled the products of this cataclysmic creation. Life did not arise with such speed, but I suspect that it originated in a tiny fraction of its subsequent duration. But the reshuffling and subsequent evolution of DNA have not simply recycled the original products; they have produced wonders.

22 | Crazy Old Randolph Kirkpatrick

OBLIVION, NOT INFAMY, is the usual fate of a crackpot. I shall be more than mildly surprised if any reader (who is not a professional taxonomist with a special attachment to sponges) can identify Randolph Kirkpatrick.

On the surface, Kirkpatrick fit the stereotype of a self-effacing, mild-mannered, dedicated, but slightly eccentric British natural historian. He was the assistant keeper of "lower" invertebrates at the British Museum from 1886 until his retirement in 1927. (I have always admired the English penchant for simple, literal terms—lifts and flats for our elevators and apartments, for example. We use the Latin *curator* for guardians of museum collections; the British call them "keepers." We, however, have done better in retaining "fall" for their "autumn.") Kirkpatrick trained as a medical student, but decided on a "less strenuous career" in natural history after several bouts with illness. He chose well, for he traveled all over the world searching for specimens and lived to be eighty-seven. In the last months of his life, in 1950, he continued to pedal his bicycle through London's busiest streets.

Early in his career, Kirkpatrick published some sound taxonomic work on sponges, but his name rarely appears in scientific journals after the First World War. In an obituary note, his successor attributed this halt in mid-career to Kirkpatrick's behavior as "an ideal public servant." "Unassuming to a fault, courteous and generous, he would spare no effort to help either a colleague or a visiting student. It was in all probability his extreme willingness to interrupt whatever he was doing to help others that prevented his completing his work."

Kirkpatrick's story, however, is by no means so simple and conventionally spotless. He did not stop publishing in 1915; instead, he shifted to private printing for a series of works that he knew no scientific journal would touch. Kirkpatrick spent the rest of his career developing what has to be the nuttiest of crackpot theories developed in this century by a professional natural historian (and keeper at the staid British Museum, no less). I do not challenge this usual assessment of his "nummulosphere" theory, but I will stoutly defend Kirkpatrick.

In 1912, Kirkpatrick was collecting sponges off the island of Porto Santo in the Madeira group, west of Morocco. One day, a friend brought him some volcanic rocks collected on a peak 1,000 feet above sea level. Kirkpatrick described his great discovery: "I examined them carefully under my binocular microscope and found to my amazement traces of nummulitic disks in all of them. Next day I visited the place whence the fragments had come."

Now *Nummulites* is one of the largest forams that ever lived (forams are single-celled creatures related to amoebas, but they secrete shells and are commonly preserved as fossils). *Nummulites* looks like the object that provided its name: a coin. Its shell is a flat disk up to an inch or two in diameter. The disk is built of individual chambers, one following the next and all wound tightly into a single coil. (The shell looks much like a coil of rope, appropriately scaled down.) Nummulites were so abundant in early Tertiary times (about 50 million years ago) that some rocks are composed almost entirely of their shells; these are called "nummulitic limestones." Nummulites litter the ground around Cairo; the Greek geographer Strabo identified them as petrified lentils left over from rations doled out to slaves who had built the Great Pyramids.

Kirkpatrick then returned to Madeira and "discovered"

nummulites in the igneous rocks there as well. I can scarcely imagine a more radical claim about the earth's structure. Igneous rocks are the products of volcanic eruption or the cooling of molten magmas within the earth; they cannot contain fossils. But Kirkpatrick argued that the igneous rocks of Madeira and Porto Santo not only included nummulites but were actually made of them. Therefore, "igneous" rocks must be sediments deposited at the ocean bottom, not the products of molten material from the earth's interior. Kirkpatrick wrote:

After the discovery of the nummulitic nature of nearly the whole island of Porto Santo, of the buildings, wine-presses, soil, etc., the name *Eozoon portosantum* seemed a fitting one for the fossils. [*Eozoon* means "dawn animal," more on it in a moment.] When the igneous rocks of Madeira were likewise found to be nummulitic, *Eozoon atlanticum* seemed a more fitting name.

Nothing could stop Kirkpatrick now. He returned to London, itching to examine igneous rocks from other areas of the world. All were made of nummulites! "I annexed in one morning for *Eozoon* volcanic rocks of the Arctic and in the afternoon of the same day those of the Pacific, Indian and Atlantic oceans. The designation *Eozoon orbis-terrarum* then suggested itself." Finally, he looked at meteorites and, yes, you guessed it, all nummulites:

If *Eozoon*, after taking in the world, had sighed for more worlds to conquer, its fortunes would have surpassed those of Alexander, for its desires would have been realized. When the empire of the nummulites was found to extend to space a final alteration of name to *Eozoon universum* apparently became necessary.

Kirkpatrick did not shy away from the evident conclusion: —all rocks on the earth's surface (including the influx from space) are made of fossils: "The original organic nature of these rocks is to me self-evident, because I can see the

Foraminiferal structure in them, and often very clearly in deed." Kirkpatrick claimed that he could see the nummu lites with a low-power hand lens, although no one eve agreed with him. "My views on igneous and certain othe rocks," he wrote, "have been received with a good deal o skepticism, and this is not surprising."

I hope I will not be dismissed as an establishment dogma tist if I state with some assurance that Kirkpatrick had some how managed to delude himself. By his own admission, h often had to work very hard in toeing his own line: "Some times I have found it necessary to examine a fragment o rock with the closest scrutiny for hours before convincing myself that I have seen all the above-mentioned details."

But what version of the earth's history would yield a crus made entirely of nummulites? Kirkpatrick proposed tha nummulites had arisen early in the history of life as the firs creatures with shells. Hence, he adopted for them the nam *Eozoon*, first proposed in the 1850s by the great Canadia geologist Sir J. W. Dawson for a supposed fossil from som of the earth's oldest rocks. (We now know that *Eozoon* is a inorganic structure, made of alternating white and gree layers of the minerals calcite and serpentine—see essay 23.

In these early times, Kirkpatrick speculated, the ocea bottom must have accumulated a deep deposit of nummu litic shells over its entire surface, for the seas contained n predators to digest them. Heat from the earth's interio fused them together and injected them with silica (thu solving the vexatious problem of why igneous rocks ar silicates, while true nummulites are made of calcium car bonate). As the nummulites were squeezed and fused, som were pushed upward and tossed out into space, later t descend as nummulitic meteorites.

Rocks are sometimes classified as fossiliferous and un- fossiliferous, but all are fossiliferous. . . . Really, then, there is, broadly speaking, one rock. . . . The litho- sphere is veritably a silicated nummulosphere.

Kirkpatrick still was not satisfied. He thought he had discovered something even more fundamental. Not content with the earth's crust and its meteorites, he began to see the coiled form of nummulites as an expression of life's essence, as the architecture of life itself. Finally, he broadened his claim to its limit: we should not say that the rocks are nummulites; rather, the rocks and the nummulites and everything else alive are expressions of "the fundamental structure of living matter," the spiral form of all existence.

Nutty, yes (unless you feel that he had intuited the double helix). Inspired, surely. A method to his madness, yes, again —and this is the crucial point. In framing his nummulosphere theory, Kirkpatrick followed the procedure that motivated all his scientific work. He had an uncritical passion for synthesis and an imagination that compelled him to gather truly disparate things together. He consistently sought similarities of geometric form among objects conventionally classified in different categories, while ignoring the ancient truth that similarity of form need not designate common cause. He also constructed similarities out of his hopes, rather than his observations.

Still, an uncautious search for synthesis may uncover real connections that would never occur to a sober scientist (although he may be jostled to reflect upon them once someone else makes the initial suggestion). Scientists like Kirkpatrick pay a heavy price, for they are usually wrong. But when they are right, they may be so outstandingly right that their insights beggar the honest work of many scientific lifetimes in conventional channels.

Let us return then to Kirkpatrick and ask why he was on Madeira and Porto Santo in the first place when he made his fateful discovery in 1912. "In September 1912," he writes, "I journeyed to Porto Santo via Madeira, in order to complete my investigation of that strange organism, the sponge-alga *Merlia normani*," In 1900, a taxonomist named J. J. Lister had discovered a peculiar sponge on the Pacific islands of Lifu and Funafuti. It contained spicules of silica, but had an additional calcareous skeleton bearing a striking resemblance to some corals (spicules are the small, needle-

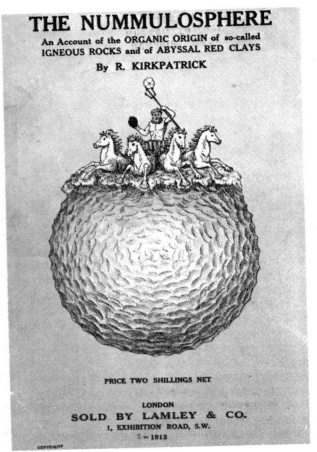

THE NUMMULOSPHERE

An Account of the ORGANIC ORIGIN of so-called
IGNEOUS ROCKS and of ABYSSAL RED CLAYS

By R. KIRKPATRICK

PRICE TWO SHILLINGS NET

LONDON
SOLD BY LAMLEY & CO.
1, EXHIBITION ROAD, S.W.
— 1913

The cover to Kirkpatrick's privately published Nummulosphere. Of it, he writes: "The design on the cover represents Neptune on the globe of waters. On one of the prongs of his trident is a piece of volcanic rock in the shape of a nummulitic disk, and in his hand is a meteorite. These emblems signify that Neptune's domain is enlarged not only at the expense of nether Jove, but also at that of high Jove whose supposed emblem of sovereignty—the thunderbolt—really belongs to the Sea God . . . Neptune's bolt is poised ready to be hurled at rash and ignorant mortals of the type of the a priori would-be refuter, daring to dispute the validity of his title-deeds."

like elements forming the skeleton of most sponges). A sober man, Lister could not accept the "hybrid" of silica and calcite; he conjectured that the spicules had entered the sponge from elsewhere. But Kirkpatrick collected more specimens and correctly concluded that the sponge secretes the spicules. Then, in 1910, Kirkpatrick found *Merlia normani* on Madeira, a second sponge with siliceous spicules and a supplementary calcareous skeleton.

Inevitably, Kirkpatrick unleashed his passion for synthesis upon *Merlia*. He noticed that its calcareous skeleton resembled several problematic groups of fossils usually classified among the corals—the stromatoporoids and the chaetetid tabulates in particular. (This may seem like a small issue to many, but I assure you that it is a major concern of all professional paleontologists. Stromatoporoids and chaetetids are very common as fossils; they form reefs in some ancient deposits. Their status lies among the classical mysteries of my field, and many distinguished paleontologists have spent entire careers devoted to their study.) Kirkpatrick decided that these and other enigmatic fossils must be sponges. He set out to find spicules in them, a sure sign of affinity with sponges. Sure enough; they all contained spicules. We may be quite sure that Kirkpatrick had deluded himself again in some cases, for he included among his "sponges" the undoubted bryozoan *Monticulipora*. In any case, Kirkpatrick soon became preoccupied with his nummulosphere theory. He never published the major treatise that he had planned on *Merlia*. The nummulosphere made him a scientific pariah, and his work on coralline sponges was pretty much forgotten.

Kirkpatrick worked the same way in studying both nummulospheres and coralline sponges: he invoked a similarity of abstract, geometric form to infer a common source for objects that no one had thought to unite, and he followed his theory with such passion that he eventually "saw" the expected form, even where it manifestly did not exist. Yet, I must note one major difference between the two studies: Kirkpatrick was right about the sponges.

During the 1960s, Thomas Goreau, late of the Discovery

Bay Marine Laboratory in Jamaica, began to explore the cryptic environments of West Indian reefs. These cracks, crevices, and caves contain a major fauna, previously undetected. In one of the most exciting zoological discoveries of the last twenty years, Goreau and his colleagues Jeremy Jackson and Willard Hartman showed that these habitats contain numerous "living fossils." This cryptic community seems to represent an entire ecosystem literally overshadowed by the evolution of more modern forms. The community may be cryptic, but its members are neither moribund nor uncommon. The linings of caves and crevices form a major part of modern reefs. Before the advent of scuba diving, scientists could not gain access to these areas.

Two elements dominate this cryptic fauna: brachiopods and Kirkpatrick's coralline sponges. Goreau and Hartman described six species of coralline sponges from the fore-reef slope of Jamaica's reef. These species form the basis for an entire new class of sponges, the Sclerospongiae. In the course of their work, they rediscovered Kirkpatrick's papers and studied his opinion on the relationship between coralline sponges and the enigmatic fossil stromatoporoids and chaetetids. "Kirkpatrick's comments," they write, "have led us to compare the coralline sponges described above with representatives of several groups of organisms known from the fossil record." They have shown, quite convincingly I think, that these fossils are indeed sponges. A major zoological discovery has solved an outstanding problem in paleontology. And crazy old Randolph Kirkpatrick had known it all along.

When I wrote to Hartman to inquire about Kirkpatrick, he cautioned me not to judge the man too harshly on his nummulosphere, for his taxonomic work on sponges had been sound. But I respect Kirkpatrick both for his sponges and for his numinous nummulosphere. It is easy to dismiss a crazy theory with laughter that debars any attempt to understand a man's motivation—and the nummulosphere is a crazy theory. I find that few men of imagination are not worth my attention. Their ideas may be wrong, even foolish, but their methods often repay a close study. Few honest passions are not based upon some valid perception of unity or some anomaly worthy of note. The different drummer often beats a fruitful tempo.

23 | Bathybius and Eozoon

WHEN THOMAS HENRY Huxley lost his young son, "our delight and our joy," to scarlet fever, Charles Kingsley tried to console him with a long peroration on the soul's immortality. Huxley, who invented the word "agnostic" to describe his own feelings, thanked Kingsley for his concern, but rejected the proffered comfort for want of evidence. In a famous passage, since taken by many scientists as a motto for proper action, he wrote: "My business is to teach my aspirations to conform themselves to fact, not to try and make facts harmonize with my aspirations. . . . Sit down before fact as a little child, be prepared to give up every preconceived notion, follow humbly wherever and to whatever abysses nature leads, or you shall learn nothing." Huxley's sentiments were noble, his grief affecting. But Huxley did not follow his own dictum, and no creative scientist ever has.

Great thinkers are never passive before facts. They ask questions of nature; they do not follow her humbly. They have hopes and hunches, and they try hard to construct the world in their light. Hence, great thinkers also make great errors.

Biologists have written a long and special chapter in the catalog of major mistakes—imaginary animals that should exist in theory. Voltaire spoke truly when he quipped: "If God did not exist, it would be necessary to invent him." Two related and intersecting chimeras arose during the early days of evolutionary theory—two animals that should have been, by Darwin's criteria, but were not. One of them had Thomas Henry Huxley for a godfather.

For most creationists, the gap between living and nonliving posed no particular problem. God had simply made the living, fully distinct and more advanced than the rocks and chemicals. Evolutionists sought to close all the gaps. Ernst Haeckel, Darwin's chief defender in Germany and surely the most speculative and imaginative of early evolutionists, constructed hypothetical organisms to span all the spaces. The lowly amoeba could not serve as a model of the earliest life, for its internal differentiation into nucleus and cytoplasm indicated a large advance from primal formlessness. Thus Haeckel proposed a lowlier organism composed only of unorganized protoplasm, the Monera. (In a way, he was right. We use his name today for the kingdom of bacteria and blue green algae, organisms without nucleus or mitochondria—although scarcely formless in Haeckel's sense.)

Haeckel defined his moneran as "an entirely homogeneous and structureless substance, a living particle of albumin, capable of nourishment and reproduction." He proposed the moneran as an intermediate form between non-living and living. He hoped that it would solve the vexing question of life's origin from the inorganic, for no problem seemed thornier for evolutionists and no issue attracted more rear-guard support for creationism than the apparent gap between the most complex chemicals and the simplest organisms. Haeckel wrote: "Every true cell already shows a division into two different parts, i.e., nucleus and plasm. The immediate production of such an object from spontaneous generation is obviously only conceivable with difficulty; but it is much easier to conceive of the production of an entirely homogeneous, organic substance, such as the structureless albumin body of the Monera."

During the 1860s, the identification of monerans assumed high priority on the agenda of Darwin's champions. And the more structureless and diffuse the moneran, the better. Huxley had told Kingsley that he would follow facts into a metaphorical abyss. But when he examined a true abyss in 1868, his hopes and expectations guided his observations. He studied some mud samples dredged from the sea bottom northwest of Ireland ten years before. He observed an inchoate, gelatinous substance in the samples. Embedded in it were tiny, circular, calcareous plates called coccoliths. Huxley identified his jelly as the heralded, formless moneran and the coccoliths as its primordial skeleton. (We now know that coccoliths are fragments of algal skeletons, which sink to the ocean bottom following the death of their planktonic producers.) Honoring Haeckel's prediction, he named it *Bathybius Haeckelii*. "I hope that you will not be ashamed of your godchild," he wrote to Haeckel. Haeckel replied that he was "very proud," and ended his note with a rallying cry: "Viva Monera."

Since nothing is quite so convincing as an anticipated discovery, *Bathybius* began to crop up everywhere. Sir Charles Wyville Thomson dredged a sample from the depths of the Atlantic and wrote: "The mud was actually alive; it stuck together in lumps, as if there were white of egg mixed with it; and the glairy mass proved, under the microscope, to be a living sarcode. Prof. Huxley . . . calls it *Bathybius.*" (The Sarcodina are a group of single-celled protozoans.) Haeckel, following his usual penchant, soon generalized and imagined that the entire ocean floor (below 5,000 feet) lay covered with a pulsating film of living *Bathybius,* the *Urschleim* (original slime) of the romantic nature philosophers (Goethe was one) idolized by Haeckel during his youth. Huxley, departing from his usual sobriety, delivered a speech in 1870 and proclaimed: "The *Bathybius* formed a living scum or film on the seabed, extending over thousands upon thousands of square miles . . . it probably forms one continuous scum of living matter girding the whole surface of the earth."

Having reached its limits of extension in space, *Bathybius* oozed out to conquer the only realm left—time. And here it met our second chimera.

Eozoon canadense, the dawn animal of Canada, was another organism whose time had come. The fossil record had caused Darwin more grief than joy. Nothing distressed him more than the Cambrian explosion, the coincident appearance of almost all complex organic designs, not near the beginning of the earth's history, but more than five-sixths of the way through it. His opponents interpreted this event as the moment of creation, for not a single trace of Precambrian life had been discovered when Darwin wrote the *Origin of Species.* (We now have an extensive record of monerans from these early rocks, see essay 21.) Nothing could have been more welcome than a Precambrian organism, the simpler and more formless the better.

In 1858, a collector for the Geological Survey of Canada found some curious specimens among the world's oldest rocks. They were made of thin, concentric layers, alternating between serpentine (a silicate) and calcium carbonate. Sir William Logan, director of the Survey, thought that they might be fossils and displayed them to various scientists, receiving in return little encouragement for his views.

Logan found some better specimens near Ottawa in 1864, and brought them to Canada's leading paleontologist, J. William Dawson, principal of McGill University. Dawson found "organic" structures, including a system of canals, in the calcite. He identified the concentric layering as the skeleton of a giant foraminifer, more diffusely formed but hundreds of times larger than any modern relative. He named it *Eozoon canadense,* the Canadian dawn animal.

Darwin was delighted. *Eozoon* entered the fourth edition of the *Origin of Species* with Darwin's firm blessing: "It is impossible to feel any doubt regarding its organic nature." (Ironically, Dawson himself was a staunch creationist, probably the last prominent holdout against evolution. As late as 1897, he wrote *Relics of Primeval Life,* a book about *Eozoon.* In it he argues that the persistence of simple Foraminifera throughout geologic time disproves natural selection since any struggle for existence would replace such lowly creatures with something more exalted.)

Bathybius and *Eozoon* were destined for union. They shared the desired property of diffuse formlessness and differed only in *Eozoon*'s discrete skeleton. Either *Eozoon* had lost its shell to become *Bathybius* or the two primordial creatures were closely related as exemplars of organic simplicity. The great physiologist W. B. Carpenter, a champion of both creatures, wrote:

> If *Bathybius* . . . could form for itself a shelly envelope, that envelope would closely resemble *Eozoon*. Further, as Prof. Huxley has proved the existence of *Bathybius* through a great range not merely of depth but of temperature, I cannot but think it probable that it has existed continuously in the deep seas of all geological epochs. . . . I am fully prepared to believe that *Eozoon*, as well as *Bathybius*, may have maintained its existence through the whole duration of geological time.

Here was a vision to titillate any evolutionist! The anticipated, formless organic matter had been found, and it extended throughout time and space to cover the floor of the mysterious and primal ocean bottom.

Before I chronicle the downfall of both creatures, I want to identify a bias that lay unstated and undefended in all the primary literature. All participants in the debate accepted without question the "obvious" truth that the most primitive life would be homogeneous and formless, diffuse and inchoate.

Carpenter wrote that *Bathybius* was "a type even lower *because less definite*, than that of Sponges." Haeckel declared that "protoplasm exists here in its simplest and earliest form, i.e., it has scarcely any definite form, and is scarcely individualized." According to Huxley, life without the internal complexity of a nucleus proved that organization arose from indefinite vitality, not vice versa: *Bathybius* "proves the absence of any mysterious power in nuclei, and shows that life is a property of the molecules of living matter, and that organization is the result of life, not life the result of organization."

But why, when we begin to think about it, should we equate formless with primitive? Modern organisms encourage no such view. Viruses are scarcely matched for regularity and repetition of form. The simplest bacteria have denite shapes. The taxonomic group that houses the amoeba, that prototype of slithering disorganization, also accommodates the Radiolaria, the most beautiful and most complexly sculpted of all regular organisms. DNA is a miracle of organization; Watson and Crick elucidated its structure by building an accurate Tinkertoy model and making sure that all the pieces fit. I would not assert any mystical Pythagorean notion that regular form underlies all organization, but I would argue that the equation of primitive with formless has roots in the outdated progressivist metaphor that views organic history as a ladder leading inexorably through all the stages of complexity from nothingness to our own noble form. Good for the ego to be sure, but not a very good outline of our world.

In any case, neither *Bathybius* nor *Eozoon* outlived Queen Victoria. The same Sir Charles Wyville Thomson who had spoken so glowingly of *Bathybius* as a "glairy mass . . . actually alive" later became chief scientist of the *Challenger* expedition during the 1870s, the most famous of all scientific voyages to explore the world's oceans. The *Challenger* scientists tried again and again to find *Bathybius* in fresh samples of deep-sea mud, but with no success.

When scientists stored mud samples for later analysis, they traditionally added alcohol to preserve organic material. Huxley's original *Bathybius* had been found in samples stored with alcohol for more than a decade. One member of the *Challenger* expedition noticed that *Bathybius* appeared whenever he added alcohol to a fresh sample. The expedition's chemist then analyzed *Bathybius* and found it to be no more than a colloidal precipitate of calcium sulfate, a product of the reaction of mud with alcohol. Thomson wrote to Huxley, and Huxley—without complaining—ate crow (or ate leeks, as he put it). Haeckel, as expected, proved more stubborn, but *Bathybius* quietly faded away.

Eozoon hung on longer. Dawson defended it literally to the death in some of the most acerbic comments ever written by a scientist. Of one German critic, he remarked in 1897: "Mobius, I have no doubt, did his best from his special and limited point of view; but it was a crime which science should not readily pardon or forget, on the part of editors of the German periodical, to publish and illustrate as scientific material a paper which was so very far from being either fair or adequate." Dawson, by that time, was a lonely holdout (although Kirkpatrick of essay 22 revived *Eozoon* in a more bizarre form later). All scientists had agreed that *Eozoon* was inorganic—a metamorphic product of heat and pressure. Indeed, it had only been found in highly metamorphosed rock, a singularly inauspicious place to find a fossil. If any more proof had been needed, the discovery of *Eozoon* in blocks of limestone ejected from Mount Vesuvius settled the issue in 1894.

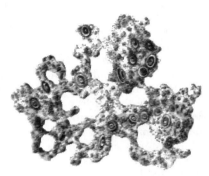

Haeckel's original illustration of *Bathybius.*
The discoidal structures are coccoliths in
the gelatinous mass.

Bathybius and *Eozoon,* ever since, have been treated by scientists as an embarrassment best forgotten. The conspiracy succeeded admirably, and I would be surprised if one percent of modern biologists ever heard of the two fantasies. Historians, trained in the older (and invalidated) tradition of science as a march to truth mediated by the successive shucking of error, also kept their peace. What can we get from errors except a good laugh or a compendium of moral homilies framed as "don'ts"?

Modern historians of science have more respect for such

BATHYBIUS AND EOZOON | 243

inspired errors. They made sense in their own time; that they don't in ours is irrelevant. Our century is no standard for all ages; science is always an interaction of prevailing culture, individual eccentricity, and empirical constraint. Hence, *Bathybius* and *Eozoon* have received more attention in the 1970s than in all previous years since their downfall. (In writing this essay, I was guided to original sources and greatly enlightened by articles of C. F. O'Brien on *Eozoon*, and N. A. Rupke and P. F. Rehbock on *Bathybius*. The article by Rehbock is particularly thorough and insightful.)

Science contains few outright fools. Errors usually have their good reasons once we penetrate their context properly and avoid judgment according to our current perception of "truth." They are usually more enlightening than embarrassing, for they are signs of changing contexts. The best thinkers have the imagination to create organizing visions, and they are sufficiently adventurous (or egotistical) to float them in a complex world that can never answer "yes" in all detail. The study of inspired error should not engender a homily about the sin of pride; it should lead us to a recognition that the capacity for great insight and great error are opposite sides of the same coin—and that the currency of both is brilliance.

Bathybius was surely an inspired error. It served the larger truth of advancing evolutionary theory. It provided a captivating vision of primordial life, extended throughout time and space. As Rehbock argues, it played a plethora of roles as, simultaneously, lowliest form of protozoology, elemental unit of cytology, evolutionary precursor of all organisms, first organic form in the fossil record, major constituent of modern marine sediments (in its coccoliths), and source of food for higher life in the nutritionally impoverished deep oceans. When *Bathybius* faded away, the problems that it had defined did not disappear. *Bathybius* inspired a great amount of fruitful scientific work and served as a focus for defining important problems still very much with us.

Orthodoxy can be as stubborn in science as in religion. I do not know how to shake it except by vigorous imagination that inspires unconventional work and contains within itself an elevated potential for inspired error. As the great Italian economist Vilfredo Pareto wrote: "Give me a fruitful error any time, full of seeds, bursting with its own corrections. You can keep your sterile truth for yourself." Not to mention a man named Thomas Henry Huxley who, when not in the throes of grief or the wars of parson hunting, argued that "irrationally held truths may be more harmful than reasoned errors."

24 | Might We Fit Inside a Sponge's Cell

I SPENT DECEMBER 31, 1979 reading through a stack of New York Sunday papers for the last weekend of the decade. Prominently featured, as always in the doldrums of such artificial transition, were lists of predictions about "ins" and "outs" across the boundary: what will the eighties reject that the seventies treasured? what, despised during the seventies, will the eighties rediscover? This surfeit of contemporary speculation drove my mind back to the last transition between centuries and to a consideration of biological ins and outs at this broader scale. The hottest subject of nineteenth-century biology did suffer a pronounced eclipse in the twentieth. Yet I happen to maintain a strong fondness for it. I also believe that new methods will revive it as a major concern for the remaining decades of our century.

Darwin's revolution led a generation of natural historians to view the reconstruction of life's tree as their most important evolutionary task. As ambitious men embarked upon a bold new course, they did not focus narrowly upon little twiglets (the relation of lions to tigers), or even upon ordinary branches (the link between cockles and mussels); they sought to root the trunk itself and to identify its major limbs: how are plants and animals related? from what source did the vertebrates spring?

In their mistaken view, these naturalists also possessed a method that could extract the answers they sought from the

spotty data at their disposal. For, under Haeckel's "bioge₁
etic law"—ontogeny recapitulates phylogeny—an anim;
climbs its own family tree during its embryological develoⱼ
ment. The simple observation of embryos should reveal ;
parade of adult ancestors in proper order. (Nothing is ev₁
quite so uncomplicated, of course. The recapitulationis
knew that some embryonic stages represented immedia₁
adaptations, not ancestral reminiscences; they also und₁
stood that stages could be mixed up, even inverted, ₁
unequal rates of development among different organs. Y
they believed that such "superficial" modifications cou
always be recognized and subtracted, leaving the ancest₁
parade intact.) E.G. Conklin, who later became an opp₁
nent of "phylogenizing," recalled the beguiling appeal ₁
Haeckel's law:

> Here was a method which promised to reveal more
> important secrets of the past than would the unearth-
> ing of all the buried monuments of antiquity—in fact
> nothing less than a complete genealogical tree of all the
> diversified forms of life which inhabit the earth.

But the turn of the century also heralded the collapse ₁
recapitulation. It died primarily because Mendelian gen₁
ics (rediscovered in 1900) rendered its premises untenab
(The "parade of adults" required that evolution proce₁
only by an addition of new stages to the end of ancest₁
ontogenies. But if new features are controlled by genes, a₁
these genes must be present from the very moment of co₁
ception, then why shouldn't new features be expressed ;
any stage of embryonic development or later growth?) B
its luster had faded long before. The assumption that anc₁
tral reminiscences could always be distinguished from r₁
cent embryonic adaptations had not been sustained. T₁
many stages were missing, too many others disco₁
bobulated. The application of Haeckel's law produced en₁
less, unresolvable, fruitless argument, not an unambiguo
tree of life. Some tree builders wanted to derive vertebra₁
from echinoderms, others from annelid worms, still oth₁

from horseshoe crabs. E.B. Wilson, apostle of the "exact," experimental method that would supplant speculative phylogenizing, complained in 1894:

> It is a ground of reproach to morphologists that their science should be burdened with such a mass of phylogenetic speculations and hypotheses, many of them mutually exclusive, in the absence of any well-defined standard of value by which to estimate their relative probability. The truth is that the search . . . has too often led to a wild speculation unworthy of the name of science; and it would be small wonder if the modern student, especially after a training in the methods of more exact sciences, should regard the whole phylogenetic aspect of morphology as a kind of speculative pedantry unworthy of serious attention.

Phylogenizing fell from general favor, but you can't keep an intrinsically exciting subject down. (I speak of high-level phylogenizing—the trunk and limbs. For twigs and small branches, where evidence is more adequate, work has always proceeded apace, with more assurance and less excitement.) We didn't need "Roots" to remind us that genealogy exerts a strange fascination over people. If uncovering the traces of a distant great-grandparent in a small overseas village fills us with satisfaction, then probing further back to an African ape, a reptile, a fish, that still-unknown ancestor of vertebrates, a single-celled forebear, even to the origin of life itself, can be positively awesome. Unfortunately, one might even say perversely, the further back we go, the more fascinated we become and the less we know. In this column, I will discuss one classic issue in phylogenizing as an example of the joys and frustrations of a subject that will not go away: the origin of multicellularity in animals.

Ideally, we might hold out for a simple, empirical resolution of the issue. Might we not hope to find a sequence of fossils so perfectly intermediate between a protist (single-celled ancestor) and a metazoan (multicelled descendant) that all doubt would be erased? We may effectively write off such a hope: the transition occurred in unfossilizable, soft-bodied creatures long before the inception of an adequate fossil record during the Cambrian explosion, some 600 million years ago. The first metazoan fossils do not surpass the most primitive modern metazoans in their similarity to protists. We must turn to living organisms, hoping that some still preserve appropriate marks of ancestry.

There is no mystery to the method of genealogical reconstruction. It is based on the analysis of similarities between postulated relatives. "Similarity," unfortunately, is no simple concept. It arises for two fundamentally different reasons. The construction of evolutionary trees requires that the two be rigorously separated, for one indicates genealogy while the other simply misleads us. Two organisms may maintain the same feature because both inherited it from a common ancestor. These are *homologous* similarities, and they indicate "propinquity of descent," to use Darwin's words. Forelimbs of people, porpoises, bats and horses provide the classic example of homology in most textbooks. They look different, and do different things, but are built of the same bones. No engineer, starting from scratch each time, would have built such disparate structures from the same parts. Therefore, the parts existed before the particular set of structures now housing them: they were, in short, inherited from a common ancestor.

Two organisms may also share a feature in common as a result of separate but similar evolutionary change in independent lineages. These are *analogous* similarities; they are the bugbear of genealogists because they confound our naive expectation that things looking alike should be closely related. The wings of birds, bats and butterflies adorn most texts as a standard example of analogy. No common ancestor of any pair had wings.

Our difficulties in identifying the trunks and limbs of life's tree do not record muddled thinking about methods. All major naturalists, from Haeckel on (and even before) stated their procedure correctly: separate homologous from analogous similarity, discard analogies, and build genealogy from homology alone. Haeckel's law was a procedure, unfortunately incorrect, for the recognition of homology. The goal is, and has been, clear enough.

In a broad sense, we know how to identify homology. Analogy has its limits. It may build striking external, functional similarity in two unrelated lineages, but it does not modify thousands of complex and independent parts in the same way. At a certain level of precision, similarities must be homologous. Unfortunately, we rarely have enough information to be confident that this required level has been attained. When we compare primitive metazoans with different protists as potential relatives, we often work with only a few features held in common for any contrast—too few to be sure about homology. Moreover, small genetic changes often have profound effects upon external, adult form. Therefore, a similarity that looks too uncanny and complex to arise more than once may actually record a simple and repeatable change. Most importantly, we aren't even comparing the right organisms, but only pale reflections of them. The transition from protist to metazoan occurred more than 600 million years ago. All true ancestors and original descendants disappeared eons ago. We can only hope that their essential, identifying features have been retained in some modern forms. Yet, if retained, they have surely been modified and overlain with a plethora of specialized adaptations. How can we separate original structure from later modification from new adaptation? No one has ever found an unfailing guide.

Only two scenarios have been favored for the origin of metazoans from protists: in the first (amalgamation) a group of protistan cells came together, began to live as a colony, evolved a division of labor and function among cells and regions, and finally formed an integrated structure; in the second (division), cellular partitions formed within a single protistan cell. (A third potential scenario, repeated failure of daughter cells to separate following cell division, has few takers these days.)

At the very outset of our inquiry, we come up against the problem of homology. What about multicellularity itself? Did it arise only once? Have we explained its occurrence in all animals once we decide how it arose in the most primitive? Or did it evolve several times? In other words, is the multicellularity of various animal lineages homologous or analogous?

The metazoan group usually regarded as most primitive, the sponges, clearly arose by the first scenario of amalgamation. In fact, modern sponges are little more than loosely knit federations of flagellated protists. In some species, cells can even be disaggregated by passing the sponge through a fine silk cloth. The cells then move independently, reaggregate into small clumps, differentiate and regenerate an entire new sponge in its original form. If all animals arose from sponges, then multicellularity is homologous throughout our kingdom, and it arose by amalgamation.

But most biologists regard sponges as an evolutionary dead end without subsequent descendants. Multicellularity is, after all, a prime candidate for frequent, independent evolution. It displays the two primary features of analogous similarity: it is reasonably simple to accomplish, and it is both highly adaptive and the only potential path to the benefits it confers. Single cells, ostrich eggs notwithstanding, cannot become very large. The earth's physical environment contains scores of habitats available only to creatures beyond the size limit of a single cell. (Consider only the stability that arises from being large enough to enter a realm where gravity overshadows the forces that act upon surfaces. Since the surface/volume ratio declines with growth, increasing size is the surest path to this realm.)

Not only has multicellularity evolved separately in the three great higher kingdoms of life (plants, animals, and fungi), but it probably arose several times in each kingdom. Most biologists agree that all origins within plants and fungi occurred by amalgamation—these organisms are the descendants of protistan colonies. Sponges also arose by amalgamation. May we then close the issue and state that multicellularity, although analogous both across and within kingdoms, evolved in the same basic way each time? Modern protists include colonial forms that display both regular arrangement of cells and incipient differentiation. Remember the *Volvox* colonies of high school biology labs? (Actually, I must confess that I don't. I attended a public high school in New York just before Sputnik went up. We had no lab at all, though it arrived in a flash just as I left.) Some volvoxes form colonies with a definite number of cells arranged in a regular manner. The cells may differ in size, and reproductive function may be confined to those at one end. Is it such a big step to a sponge?

Only among animals may we make a good case for another scenario. Did some animals, ourselves included, arise by division? This question cannot be answered until we resolve one of the oldest riddles in zoology: the status of the phylum Cnidaria (corals and their allies, but also including the beautiful, translucent Ctenophora, or comb-jellies). Almost everyone agrees that the Cnidaria arose by amalgamation. The dilemma resides in their relationship with other animal phyla. Almost all possible schemes have their supporters: cnidarians as descendants of sponges and ancestors of nothing else; cnidarians as a separate branch of the animal kingdom without descendants; cnidarians as the ancestors of all "higher" animal phyla (the classical view of the nineteenth century); cnidarians as degenerate descendants of a higher phylum. If either of the last two schemes can ever be established, then our issue is settled—all animals arose by amalgamation, probably twice (sponges and everything else). But if the "higher" animal phyla are not closely related to cnidarians, if they represent a third, separate evolution of multicellularity in the animal kingdom, then the scenario of division must be seriously considered.

Supporters of a separate origin for the higher animals generally cite the Platyhelminthes (flatworms) as a potentially ancestral stock. Earl Hanson, a biologist at Wesleyan University, has been a leading crusader, both for a platyhelminth origin of higher animals and for the scenario of division. If his iconoclastic view prevails, then the higher animals, including humans of course, are probably the only multicellular products of division rather than amalgamation.

Hanson has pursued his case by studying the similarities

between a group of protists known as ciliates (including the familiar *Paramecium*), and the "simplest" of flatworms, the Acoela (named for their failure to develop a body cavity). Many ciliates maintain large numbers of nuclei within their single cell. If cellular partitions arose between the nuclei, would the resulting creature be enough like an acoelous flatworm to justify a claim for homology?

Hanson documents an extensive set of similarities between the multinucleate ciliates and the acoeles. Acoeles are tiny marine flatworms. Some can swim, and a few live in water up to 250 meters in depth; but most crawl along the sea bottom in shallow water, living under rocks or in sand and mud. They are similar in size to the multinucleate ciliates. (It is not true that all metazoans are larger than all protists. The ciliates range in length from 1/100 to 3 millimeters, while some acoeles are less than 1 millimeter in length.) The internal similarities of ciliates and acoeles reside primarily in their shared simplicity; for acoeles, unlike conventional metazoans, lack both a body cavity and the organs associated with it. They have no permanent digestive, excretory, or respiratory system. Like the ciliate protists, they form temporary food vacuoles and perform digestion within them. Both ciliates and acoeles divide their bodies roughly into inner and outer layers. Ciliates maintain an ectoplasm (outer layer) and endoplasm (inner layer), and concentrate their nuclei in the endoplasm. Acoeles devote an inner region to digestion and reproduction, and an outer region to locomotion, protection, and capture of food.

The two groups also display some outstanding differences. Acoeles build a nerve net and reproductive organs that can become quite complex. Some have penises, for example, and impregnate each other hypodermically by penetrating through the body wall. They undergo embryonic development after fertilization. Ciliates, by contrast, have no organized nervous system. They divide by fission and have no embryology, although they do indulge in sex via a process called conjugation. (In conjugation, two ciliates come together and exchange genetic material. They then separate and each divides later to form two daughters. Sex and reproduction, combined in nearly all metazoa, are separate processes in ciliates.) Most prominently, of course, acoeles are cellularized, ciliates are not.

These differences should not debar a hypothesis of close genealogical relationship. After all, as I argued previously, contemporary ciliates and acoeles are more than half a billion years beyond their potential common ancestor. Neither represents a transitional form in the origin of multicellularity. The debate centers instead on the similarities, and on the oldest and most basic issue of all: are the similarities homologous or analogous?

Hanson argues for homology, claiming that acoele simplicity is an ancestral condition within the platyhelminths—and that similarities between ciliates and acoeles, largely a result of this simplicity, do record genealogical connection. His detractors reply that the simplicity of acoeles is a secondary result of their "regressive" evolution from more complex platyhelminths, a consequence of pronounced reduction in body size within acoeles. Larger turbellarians (the platyhelminth group including acoeles) have intestines and excretory organs. If acoele simplicity is a derived condition *within* the turbellarians, then it cannot reflect direct inheritance from a ciliate stock.

Unfortunately, the similarities that Hanson cites are of the sort that always produce unresolvable wrangling about homology vs. analogy. They are neither precise, nor numerous enough to guarantee homology. Many are based upon the *absence* of complexity in acoeles, and evolutionary loss is easy and repeatable, whereas separate development of precise and intricate structures may be unlikely. Moreover, acoele simplicity is a predictable result of their small body size—it may represent a functional convergence upon ciliate design by a group that secondarily entered their range of body size, not a connection by descent. Again, we invoke the principle of surfaces and volumes. Many physiological functions, including breathing, digestion, and excretion, must proceed through surfaces and serve the entire body's volume. Large animals have such a low ratio of external surface to internal volume that they must evolve internal organs to provide more surface. (Functionally, lungs are little more than bags of surface for exchange of gases, while intestines are sheets of surface for the passage of digested food.) But small animals maintain such a high ratio of external surface to internal volume that they often can breathe, feed, and excrete through the external surface alone. The smallest representatives of many phyla more complex than platyhelminths also lose internal organs. *Caecum*, for example, the smallest snail, has lost its internal respiratory system entirely and takes in oxygen through its external surface.

Other similarities, cited by Hanson, may be homologous, but so widespread among other creatures that they merely illustrate the broader affinity of all protists with all metazoans, not any specific pathway of descent. Meaningful homologies must be confined to characters that are both shared by descent *and* derived. (Derived characters evolve uniquely in the common ancestor of two groups that share them; they are marks of genealogy. A shared primitive character, on the other hand, cannot specify descent. The presence of DNA in both ciliates and acoeles tells us nothing about their affinity because all protists and metazoans have DNA.) Thus, Hanson mentions "complete ciliation" as a "permanent character significantly held in common by ciliates and acoeles." But cilia, although homologous, are a shared primitive character; many other groups, including cnidarians, have them. The *completeness* of ciliation, on the other hand, represents an "easy" evolutionary event that may only be analogous in ciliates and acoeles. The external surface sets a limit to the maximal number of cilia that may be affixed. Small animals, with high surface/volume ratios, may indulge in ciliary locomotion; large animals cannot insert enough cilia on their relatively declining surface to propel their mass. The complete ciliation of acoeles may reflect a secondary, adaptive response to their small size. The tiny snail *Caecum* also moves by cilia; all its larger relatives use muscular contraction for locomotion.

Hanson is, of course, well aware that he cannot prove his

intriguing hypothesis with the classical evidence of morphology and function. "The best we can say," he concludes, "is that many suggestive similarities are present [between ciliates and acoeles], but no rigorously definable homologies." Is there another method that might resolve the issue, or are we permanently condemned to unresolvable wrangling? Homology might be established with confidence if we could generate a new set of characters sufficiently numerous, comparable, and complex—for analogy cannot be the explanation of detailed, part-by-part similarity in thousands of independent items. The laws of mathematical probability will not allow it.

Fortunately, we now have a potential source of such information—the DNA sequence of comparable proteins. All protists and metazoans share many homologous proteins. Each protein is built of a long chain of amino acids; each amino acid is coded by a sequence of three nucleotides in DNA. Thus, the DNA code for each protein may contain hundreds of thousands of nucleotides in a definite order.

Evolution proceeds by substitution of nucleotides. After two groups split from a common ancestor, their nucleotide sequences begin to accumulate changes. The number of changes seems to be at least roughly proportional to the amount of time since the split. Thus, overall similarity in nucleotide sequence for homologous proteins may measure the extent of genealogical separation. A nucleotide sequence is a homologizer's dream—for it represents thousands of potentially independent characters. Each nucleotide position is a site of possible change.

Techniques are just now becoming available for the routine sequencing of nucleotides. Within ten years, I believe, we will be able to take homologous proteins from all the ciliate and metazoan groups at issue, sequence them, measure the similarities between each pair of organisms and obtain greater insight (perhaps even resolution) for this old genealogical mystery. If acoeles are most similar to protist groups that might achieve multicellularity by evolving cell membranes within their bodies, then Hanson will be vindicated. But if they are closest to protists that can reach multicellularity by integration within a colony, then the classical view will prevail, and all metazoa will emerge as the products of amalgamation.

The study of genealogy has been unfairly eclipsed in our century by the analysis of adaptation, but it cannot lose its power to fascinate. Simply consider what Hanson's scenario implies about our relationship with other multicellular organisms. Few zoologists doubt that all higher animals achieved their multicellular status by whatever method the flatworms followed. If acoeles evolved by the cellularization of a ciliate, then our multicellular body is the homolog of a single protistan cell. If sponges, cnidarians, plants and fungi arose by amalgamation, then their bodies are the homologs of a protistan colony. Since each ciliate cell is the homolog of an individual cell in any protistan colony, we must conclude—and I do mean this literally—that the entire human body is the homolog of a single cell in a sponge, coral, or plant.

The curious paths of homology go further back. The protistan cell itself may have evolved from a symbiosis of several simpler prokaryotic (bacterial or blue green algal) cells. Mitochondria and chloroplasts seem to be the homologs of entire prokaryotic cells. Thus, each cell of any protist, and each cell in any metazoan body, may be, by genealogy, an integrated colony of prokaryotes. Shall we then view ourselves both as a congeries of bacterial colonies and as the homolog of a single cell in a sponge or onion skin? Think upon it next time you swallow a carrot or slice a mushroom.

7 | They Were Despised and Rejected

25 | Were Dinosaurs Dumb?

WHEN MUHAMMAD ALI flunked his army intelligence test, he quipped (with a wit that belied his performance on the exam): "I only said I was the greatest; I never said I was the smartest." In our metaphors and fairy tales, size and power are almost always balanced by a want of intelligence. Cunning is the refuge of the little guy. Think of Br'er Rabbit and Br'er Bear; David smiting Goliath with a slingshot; Jack chopping down the beanstalk. Slow wit is the tragic flaw of a giant.

The discovery of dinosaurs in the nineteenth century provided, or so it appeared, a quintessential case for the negative correlation of size and smarts. With their pea brains and giant bodies, dinosaurs became a symbol of lumbering stupidity. Their extinction seemed only to confirm their flawed design.

Dinosaurs were not even granted the usual solace of a giant—great physical prowess. God maintained a discreet silence about the brains of behemoth, but he certainly marveled at its strength: "Lo, now, his strength is in his loins, and his force is in the navel of his belly. He moveth his tail like a cedar. . . . His bones are as strong pieces of brass; his bones are like bars of iron [Job 40:16–18]." Dinosaurs, on the other hand, have usually been reconstructed as slow and clumsy. In the standard illustration, *Brontosaurus* wades in a murky pond because he cannot hold up his own weight on land.

Popularizations for grade school curricula provide a good illustration of prevailing orthodoxy. I still have my third grade copy (1948 edition) of Bertha Morris Parker's *Animals of Yesterday,* stolen, I am forced to suppose, from P.S. 26, Queens (sorry Mrs. McInerney). In it, boy (teleported back to the Jurassic) meets brontosaur:

> It is huge, and you can tell from the size of its head that it must be stupid. . . . This giant animal moves about very slowly as it eats. No wonder it moves slowly! Its huge feet are very heavy, and its great tail is not easy to pull around. You are not surprised that the thunder lizard likes to stay in the water so that the water will help it hold up its huge body. . . . Giant dinosaurs were once the lords of the earth. Why did they disappear? You can probably guess part of the answer—their bodies were too large for their brains. If their bodies had been smaller, and their brains larger, they might have lived on.

Dinosaurs have been making a strong comeback of late, in this age of "I'm OK, you're OK." Most paleontologists are now willing to view them as energetic, active, and capable animals. The *Brontosaurus* that wallowed in its pond a generation ago is now running on land, while pairs of males have been seen twining their necks about each other in elaborate sexual combat for access to females (much like the neck wrestling of giraffes). Modern anatomical reconstructions indicate strength and agility, and many paleontologists now believe that dinosaurs were warmblooded (see essay 26).

The idea of warmblooded dinosaurs has captured the public imagination and received a torrent of press coverage. Yet another vindication of dinosaurian capability has received very little attention, although I regard it as equally significant. I refer to the issue of stupidity and its correlation with size. The revisionist interpretation, which I support in this column, does not enshrine dinosaurs as paragons of intellect, but it does maintain that they were not

Triceratops   GREGORY S. PAUL

small brained after all. They had the "right-sized" brains for reptiles of their body size.

I don't wish to deny that the flattened, minuscule head of largebodied *Stegosaurus* houses little brain from our subjective, top-heavy perspective, but I do wish to assert that we should not expect more of the beast. First of all, large animals have relatively smaller brains than related, small animals. The correlation of brain size with body size among kindred animals (all reptiles, all mammals, for example) is remarkably regular. As we move from small to large animals, from mice to elephants or small lizards to Komodo dragons, brain size increases, but not so fast as body size. In other words, bodies grow faster than brains, and large animals have low ratios of brain weight to body weight. In fact, brains grow only about two-thirds as fast as bodies. Since we have no reason to believe that large animals are consistently stupider than their smaller relatives, we must conclude that large animals require relatively less brain to do as well as smaller animals. If we do not recognize this relationship, we are likely to underestimate the mental

Brachiosaurus   GREGORY S. PAUL

power of very large animals, dinosaurs in particular.

Second, the relationship between brain and body size is not identical in all groups of vertebrates. All share the same rate of relative decrease in brain size, but small mammals have much larger brains than small reptiles of the same body weight. This discrepancy is maintained at all larger body weights, since brain size increases at the same rate in both groups—two-thirds as fast as body size.

Put these two facts together—all large animals have relatively small brains, and reptiles have much smaller brains than mammals at any common body weight—and what should we expect from a normal, large reptile? The answer, of course, is a brain of very modest size. No living reptile even approaches a middle-sized dinosaur in bulk, so we have no modern standard to serve as a model for dinosaurs.

Fortunately, our imperfect fossil record has, for once, not

severely disappointed us in providing data about fossil brains. Superbly preserved skulls have been found for many species of dinosaurs, and cranial capacities can be measured. (Since brains do not fill craniums in reptiles, some creative, although not unreasonable, manipulation must be applied to estimate brain size from the hole within a skull.) With these data, we have a clear test for the conventional hypothesis of dinosaurian stupidity. We should agree, at the outset, that a reptilian standard is the only proper one—it is surely irrelevant that dinosaurs had smaller brains than people or whales. We have abundant data on the relationship of brain and body size in modern reptiles. Since we know that brains increase two-thirds as fast as bodies as we move from small to large living species, we can extrapolate this rate to dinosaurian sizes and ask whether dinosaur brains match what we would expect of living reptiles if they grew so large.

Harry Jerison studied the brain sizes of ten dinosaurs and found that they fell right on the extrapolated reptilian curve. Dinosaurs did not have small brains; they maintained just the right-sized brains for reptiles of their dimensions. So much for Ms. Parker's explanation of their demise.

Jerison made no attempt to distinguish among various kinds of dinosaurs; ten species distributed over six major groups scarcely provide a proper basis for comparison. Recently, James A. Hopson of the University of Chicago gathered more data and made a remarkable and satisfying discovery.

Hopson needed a common scale for all dinosaurs. He therefore compared each dinosaur brain with the average reptilian brain we would expect at its body weight. If the dinosaur falls on the standard reptilian curve, its brain receives a value of 1.0 (called an encephalization quotient, or EQ—the ratio of actual brain to expected brain for a standard reptile of the same body weight). Dinosaurs lying above the curve (more brain than expected in a standard reptile of the same body weight) receive values in excess of 1.0, while those below the curve measure less than 1.0.

Hopson found that the major groups of dinosaurs can be

ranked by increasing values of average EQ. This ranking corresponds perfectly with inferred speed, agility and behavioral complexity in feeding (or avoiding the prospect of becoming a meal). The giant sauropods, *Brontosaurus* and its allies, have the lowest EQ's—0.20 to 0.35. They must have moved fairly slowly and without great maneuverability. They probably escaped predation by virtue of their bulk alone, much as elephants do today. The armored ankylosaurs and stegosaurs come next with EQ's of 0.52 to 0.56. These animals, with their heavy armor, probably relied largely upon passive defense, but the clubbed tail of ankylosaurs and the spiked tail of stegosaurs imply some active fighting and increased behavioral complexity.

The ceratopsians rank next at about 0.7 to 0.9. Hopson remarks: "The larger ceratopsians, with their great horned heads, relied on active defensive strategies and presumably required somewhat greater agility than the tail-weaponed forms, both in fending off predators and in intraspecific combat bouts. The smaller ceratopsians, lacking true horns, would have relied on sensory acuity and speed to escape from predators." The ornithopods (duckbills and their allies) were the brainiest herbivores, with EQ's from 0.85 to 1.5. They relied upon "acute senses and relatively fast speeds" to elude carnivores. Flight seems to require more acuity and agility than standing defense. Among ceratopsians, small, hornless, and presumably fleeing *Protoceratops* had a higher EQ than great three-horned *Triceratops*.

Carnivores have higher EQ's than herbivores, as in modern vertebrates. Catching a rapidly moving or stoutly fighting prey demands a good deal more upstairs than plucking the right kind of plant. The giant theropods (*Tyrannosaurus* and its allies) vary from 1.0 to nearly 2.0. Atop the heap, quite appropriately at its small size, rests the little coelurosaur *Stenonychosaurus* with an EQ well above 5.0. Its actively moving quarry, small mammals and birds perhaps, probably posed a greater challenge in discovery and capture than *Triceratops* afforded *Tyrannosaurus*.

I do not wish to make a naive claim that brain size equals intelligence or, in this case, behavioral range and agility (I don't know what intelligence means in humans, much less in a group of extinct reptiles). Variation in brain size within a species has precious little to do with brain power (humans do equally well with 900 or 2,500 cubic centimeters of brain). But comparison across species, when the differences are large, seems reasonable. I do not regard it as irrelevant to our achievements that we so greatly exceed koala bears —much as I love them—in EQ. The sensible ordering among dinosaurs also indicates that even so coarse a measure as brain size counts for something.

If behavioral complexity is one consequence of mental power, then we might expect to uncover among dinosaurs some signs of social behavior that demand coordination, cohesiveness, and recognition. Indeed we do, and it cannot be accidental that these signs were overlooked when dinosaurs labored under the burden of a falsely imposed obtuseness. Multiple trackways have been uncovered, with evidence for more than twenty animals traveling together in parallel movement. Did some dinosaurs live in herds? At the Davenport Ranch sauropod trackway, small footprints lie in the center and larger ones at the periphery. Could it be that some dinosaurs traveled much as some advanced herbivorous mammals do today, with large adults at the borders sheltering juveniles in the center?

In addition, the very structures that seemed most bizarre and useless to older paleontologists—the elaborate crests of hadrosaurs, the frills and horns of ceratopsians, and the nine inches of solid bone above the brain of *Pachycephalosaurus*—now appear to gain a coordinated explanation as devices for sexual display and combat. Pachycephalosaurs may have engaged in head-butting contests much as mountain sheep do today. The crests of some hadrosaurs are well designed as resonating chambers; did they engage in bellowing matches? The ceratopsian horn and frill may have acted as sword and shield in the battle for mates. Since such behavior is not only intrinsically complex, but also implies an elaborate social system, we would scarcely expect to find it in a group of animals barely muddling through at a moronic level.

But the best illustration of dinosaurian capability may well be the fact most often cited against them—their demise. Extinction, for most people, carries many of the connotations attributed to sex not so long ago—a rather disreputable business, frequent in occurrence, but not to anyone's credit, and certainly not to be discussed in proper circles. But, like sex, extinction is an ineluctable part of life. It is the ultimate fate of all species, not the lot of unfortunate and ill-designed creatures. It is no sign of failure.

The remarkable thing about dinosaurs is not that they became extinct, but that they dominated the earth for so long. Dinosaurs held sway for 100 million years while mammals, all the while, lived as small animals in the interstices of their world. After 70 million years on top, we mammals have an excellent track record and good prospects for the future, but we have yet to display the staying power of dinosaurs.

People, on this criterion, are scarcely worth mentioning —5 million years perhaps since *Australopithecus*, a mere 50,-000 for our own species, *Homo sapiens*. Try the ultimate test within our system of values: Do you know anyone who would wager a substantial sum, even at favorable odds, on the proposition that *Homo sapiens* will last longer than *Brontosaurus*?

26 | The Telltale Wishbone

WHEN I WAS four I wanted to be a garbageman. I loved the rattling of the cans and the whir of the compressor; I thought that all of New York's trash might be squeezed into a single, capacious truck. Then, when I was five, my father took me to see the *Tyrannosaurus* at the American Museum of Natural History. As we stood in front of the beast, a man sneezed; I gulped and prepared to utter my *Shema Yisrael.* But the great animal stood immobile in all its bony grandeur, and as we left, I announced that I would be a paleontologist when I grew up.

In those distant days of the late 1940s, there wasn't much to nurture a boy's interest in paleontology. I remember *Fantasia,* Alley Oop, and some fake-antique metal statues in the Museum shop, priced way above my means and not very attractive anyway. Most of all, I recall the impression conveyed in books: *Brontosaurus,* wallowing its life away in ponds because it couldn't support its weight on dry land; *Tyrannosaurus,* fierce in battle but clumsy and ungainly in motion. In short, slow, lumbering, pea-brained, cold-blooded brutes. And, as the ultimate proof of their archaic insufficiency, did they not all perish in the great Cretaceous extinction?

One aspect of this conventional wisdom always bothered me: why had these deficient dinosaurs done so well—and for so long? Therapsid reptiles, the ancestors of mammals, had become diverse and abundant before the rise of the dinosaurs. Why didn't they, rather than dinosaurs, inherit the earth? Mammals themselves had evolved at about the same time as dinosaurs and had lived for 100 million years as small and uncommon creatures. Why, if dinosaurs were so slow, stupid, and inefficient, did mammals not prevail right away?

A striking resolution has been suggested by several paleontologists during the past decade. Dinosaurs, they argue, were fleet, active, and warmblooded. Moreover, they have not yet gone the way of all flesh, for a branch of their lineage persists in the branches—we call them birds.

I once vowed that I would not write about warmblooded dinosaurs in these essays: the new gospel had gone forth quite adequately in television, newspapers, magazines, and popular books. The intelligent layperson, that worthy abstraction for whom we write, must be saturated. But I relent, I think, for good reason. In nearly endless discussions, I find that the relationship between two central claims—dinosaur endothermy (warmbloodedness) and dinosaurian ancestry of birds—has been widely misunderstood. I also find that the relationship between dinosaurs and birds has provoked public excitement for the wrong reason, while the right reason, usually unappreciated, neatly unites the ancestry of birds with endothermy of dinosaurs. And this union supports the most radical proposal of all—a restructuring of vertebrate classification that removes dinosaurs from Reptilia, sinks the traditional class Aves (birds), and designates a new class, Dinosauria, uniting birds and dinosaurs. Terrestrial vertebrates would fit into four classes: two coldblooded, Amphibia and Reptilia, and two warmblooded, Dinosauria and Mammalia. I have not made up my own mind about this new classification, but I appreciate the originality and appeal of the argument.

The claim that birds had dinosaurs as ancestors is not so tumultuous as it might first appear. It involves no more than a slight reorientation of a branch on the phyletic tree. The very close relationship between *Archaeopteryx,* the first bird, and a group of small dinosaurs called coelurosaurs has never been doubted. Thomas Henry Huxley and most nineteenth-century paleontologists advocated a relationship of direct descent and derived birds from dinosaurs.

But Huxley's opinion fell into disfavor during this century for a simple, and apparently valid, reason. Complex structures, once totally lost in evolution, do not reappear in the same form. This statement invokes no mysterious directional force in evolution, but merely asserts a claim based upon mathematical probability. Complex parts are built by many genes, interacting in complex ways with the entire developmental machinery of an organism. If dismantled by evolution, how could such a system be built again, piece by piece? The rejection of Huxley's argument hinged upon a single bone—the clavicle, or collarbone. In birds, including *Archaeopteryx,* the clavicles are fused to form a furcula, better known to friends of Colonel Sanders as a wishbone. All dinosaurs, it appeared, had lost their clavicles; hence, they

Archaeopteryx   GREGORY S. PAUL

could not be the direct ancestors of birds. An unimpeachable argument if true. But negative evidence is notoriously prone to invalidation by later discovery.

Still, even Huxley's opponents could not deny the detailed structural similarity between *Archaeopteryx* and the coelurosaurian dinosaurs. So they opted for the nearest possible relationship between birds and dinosaurs—common derivation from a group of reptiles that still possessed a clavicle, subsequently lost in one line of descent (dinosaurs) and strengthened and fused in another (birds). The best candidates for common ancestry are a group of Triassic thecodont reptiles called pseudosuchians.

Many people, on first hearing that birds might be surviving dinosaurs, think that such a striking claim must represent a complete discombobulation of received doctrine about vertebrate relationships. Nothing could be further from the truth. All paleontologists advocate a close affinity between dinosaurs and birds. The current debate centers about a small shift in phyletic branching points: birds either branched from pseudosuchians or from the descendants of pseudosuchians—the coelurosaurian dinosaurs. If birds branched at the pseudosuchian level, they cannot be labeled as descendants of dinosaurs (since dinosaurs had not yet arisen); if they evolved from coelurosaurs, they are the only surviving branch from a dinosaur stem. Since pseudosuchians and primitive dinosaurs looked so much alike, the actual point of branching need not imply much about the biology of birds. No one is suggesting that hummingbirds evolved from *Stegosaurus* or *Triceratops*.

The issue, thus explicated, may now seem rather ho-hum to many readers, although I shall soon argue (for a different reason) that it isn't. But I want to emphasize that these twists of genealogy are of utmost concern to professional paleontologists. We care very much about who branched from whom because reconstructing the history of life is our business, and we value our favorite creatures with the same loving concern that most people invest in their families. Most people would care very much if they learned that their cousin was really their father—even if the discovery provided few insights about their biological construction.

Yale paleontologist John Ostrom has recently revived the dinosaurian theory. He restudied every specimen of *Archaeopteryx*—all five of them. First of all, the main objection to dinosaurs as ancestors had already been countered. At least two coelurosaurian dinosaurs had clavicles after all; they are no longer debarred as progenitors of birds. Secondly, Ostrom documents in impressive detail the extreme similarity in structure between *Archaeopteryx* and coelurosaurs. Since many of these common features are not shared by pseudosuchians, they either evolved twice (if pseudosuchians are ancestors of both birds and dinosaurs) or they evolved just once and birds inherited them from dinosaur ancestors.

Separate development of similar features is very common in evolution; we refer to it as parallelism, or convergence. We anticipate convergence in a few relatively simple and clearly adaptive structures when two groups share the same mode of life—consider the saber-toothed marsupial carnivore of South America and the placental saber-toothed "tiger" (see essay 28). But when we find part-by-part correspondence for minutiae of structure without clear adaptive necessity, then we conclude that the two groups share their similarities by descent from a common ancestor. Therefore, I accept Ostrom's revival. The only major impediment to dinosaurs as ancestors of birds had already been removed with the discovery of clavicles in some coelurosaurian dinosaurs.

Birds evolved from dinosaurs, but does this mean, to cite the litany of some popular accounts, that dinosaurs are still alive? Or, to put the question more operationally, shall we classify dinosaurs and birds in the same group, with birds as the only living representatives? Paleontologists R. T. Bakker and P. M. Galton advocated this course when they proposed the new vertebrate class Dinosauria to accommodate both birds and dinosaurs.

A decision on this question involves a basic issue in taxonomic philosophy. (Sorry to be so technical about such a hot subject, but severe misunderstandings can arise when

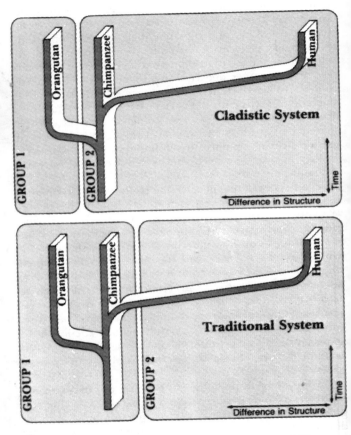

The Telltale Wishbone. With permission from *Natural History*, November, 1977. © American Museum of Natural History, 1977

we fail to sort formal questions in taxonomy from biological claims about structure and physiology.) Some taxonomists argue that we should group organisms only by patterns of branching: if two groups branch from each other and have no descendants (like dinosaurs and birds), they must be united in formal classification before either group joins another (like dinosaurs with other reptiles). In this so-called cladistic (or branching) system of taxonomy, dinosaurs cannot be reptiles unless birds are as well. And if birds are not reptiles, then according to the rules, dinosaurs and birds must form a single, new class.

Other taxonomists argue that branching points are not the only criterion of classification. They weigh the degree of adaptive divergence in structure as well. In the cladistic system, cows and lungfishes have a closer affinity than lungfishes and salmon because the ancestors of terrestrial vertebrates branched from the sarcopterygian fishes (a group including lungfishes) after the sarcopts had already branched from the actinopterygian fishes (standard bony fishes, including salmon). In the traditional system, we consider biological structure as well as branching pattern, and we may continue to classify lungfishes and salmon together as fish because they share so many common features of aquatic vertebrates. The ancestors of cows experienced an enormous evolutionary transformation, from amphibian to reptile to mammal; lungfish stagnated and look pretty much as they did 250 million years ago. Fish is fish, as an eminent philosopher once said.

The traditional system recognizes unequal evolutionary rates after branching as a proper criterion of classification. A group may win separate status by virtue of its profound divergence. Thus, in the traditional system, mammals can be a separate group and lungfishes can be kept with other fish. Humans can be a separate group and chimps can be kept with orangutans (even though humans and chimps share a more recent branching point than chimps and orangs). Similarly, birds can be a separate group and dinosaurs kept with reptiles, even though birds branched from dinosaurs. If birds developed the structural basis of their great success after they branched from dinosaurs, and if dinosaurs never diverged far from a basic reptilian plan, then birds should be grouped separately and dinosaurs kept with reptiles, despite their genealogical history of branching.

Thus, we finally arrive at our central question and at the union of this technical issue in taxonomy with the theme of warmblooded dinosaurs. Did birds inherit their primary features directly from dinosaurs? If they did, Bakker and Galton's class Dinosauria should probably be accepted, despite the adherence of most modern birds to a mode of life (flight and small size) not wonderfully close to that of most dinosaurs. After all, bats, whales, and armadillos are all mammals.

Consider the two cardinal features that provided an adaptive basis for flight in birds—feathers for lift and propulsion and warmbloodedness for maintaining the consistently high levels of metabolism required by so strenuous an activity as flight. Could *Archaeopteryx* have inherited both these features from dinosaur ancestors?

R. T. Bakker has presented the most elegant brief for warmblooded dinosaurs. He rests his controversial case on four major arguments:

1. The structure of bone. Coldblooded animals cannot keep their body temperature at a constant level: it fluctuates in sympathy with temperatures in the outside environment. Consequently, coldblooded animals living in regions with intense seasonality (cold winters and hot summers) develop growth rings in outer layers of compact bone—alternating layers of rapid summer and slower winter growth. (Tree rings, of course, record the same pattern.) Warmblooded animals do not develop rings because their internal temperature is constant in all seasons. Dinosaurs from regions of intense seasonality do not have growth rings in their bones.

2. Geographic distribution. Large coldblooded animals do not live at high latitudes (far from the equator) because they cannot warm up enough during short winter days and are too large to find safe places for hibernation. Some large dinosaurs lived so far north that they had to endure long

periods entirely devoid of sunlight during the winter.

3. Fossil ecology. Warmblooded carnivores must eat much more than coldblooded carnivores of the same size in order to maintain their constant body temperatures. Consequently, when predators and prey are about the same size, a community of coldblooded animals will include relatively more predators (since each one needs to eat so much less) than a community of warmblooded animals. The ratio of predators to prey may reach 40 percent in coldblooded communities; it does not exceed 3 percent in warmblooded communities. Predators are rare in dinosaur faunas; their relative abundance matches our expectation for modern communities of warmblooded animals.

4. Dinosaur anatomy. Dinosaurs are usually depicted as slow, lumbering beasts, but newer reconstructions (see essay 25) indicate that many large dinosaurs resembled modern running mammals in locomotor anatomy and the proportions of their limbs.

But how can we view feathers as an inheritance from dinosaurs; surely no *Brontosaurus* was ever invested like a peacock. For what did *Archaeopteryx* use its feathers? If for flight, then feathers may belong to birds alone; no one has ever postulated an airborne dinosaur (flying pterosaurs belong to a separate group). But Ostrom's anatomical reconstruction strongly suggests that *Archaeopteryx* could not fly; its feathered forearms are joined to its shoulder girdle in a manner quite inappropriate for flapping a wing. Ostrom suggests a dual function for feathers: insulation to protect a small warmblooded creature from heat loss and as a sort of basket trap to catch flying insects and other small prey in a fully enclosed embrace.

Archaeopteryx was a small animal. It weighed less than a pound, and stood a full foot shorter than the smallest dinosaur. Small creatures have a very high ratio of surface area to volume (see essays 29 and 30). Heat is generated over a body's volume and radiated out through its surface. Small warmblooded creatures have special problems in maintaining a constant body temperature since heat dissipates so quickly from their relatively enormous surface. Shrews, although insulated by a coat of hair, must eat nearly all the time to keep their internal fires burning. The ratio of surface to volume was so low in large dinosaurs that they could maintain constant temperatures without insulation. But as soon as any dinosaur or its descendant became very small, it would need insulation to remain warmblooded. Feathers may have served as a primary adaptation for constant temperatures in small dinosaurs. Bakker suggests that many small coelurosaurs may have been feathered as well. (Very few fossils preserve any feathers; *Archaeopteryx* is a great rarity of exquisite preservation.)

Feathers, evolved primarily for insulation, were soon exploited for another purpose in flight. Indeed, it is hard to imagine how feathers could have evolved if they never had a use apart from flight. The ancestors of birds were surely flightless, and feathers did not arise all at once and fully formed. How could natural selection build an adaptation through several intermediate stages in ancestors that had no use for it? By postulating a primary function for insulation, we may view feathers as a device for giving warmblooded dinosaurs an access to the ecological advantages of small size.

Ostrom's arguments for a descent of birds from coelurosaurian dinosaurs do not depend upon the warmbloodedness of dinosaurs or the primary utility of feathers as insulation. They are based instead upon the classical methods of comparative anatomy—detailed part-by-part similarity between bones, and a contention that such striking resemblance must reflect common descent, not convergence. I believe that Ostrom's arguments will stand no matter how the hot debate about warmblooded dinosaurs eventually resolves itself.

But the descent of birds from dinosaurs wins its fascination in the public eye only if birds inherited their primary adaptations of feathers and warmbloodedness directly from dinosaurs. If birds developed these adaptations after they branched, then dinosaurs are perfectly good reptiles in their physiology; they should be kept with turtles, lizards,

THE TELLTALE WISHBONE | 277

and their kin in the class Reptilia. (I tend to be a traditionalist rather than a cladist in my taxonomic philosophy.) But if dinosaurs really were warmblooded, and if feathers were their way of remaining warmblooded at small sizes, then birds inherited the basis of their success from dinosaurs. And if dinosaurs were closer to birds than to other reptiles in their physiology, then we have a classical structural argument—not just a genealogical claim—for the formal alliance of birds and dinosaurs in a new class, Dinosauria.

Bakker and Galton write: "The avian radiation is an aerial exploitation of basic dinosaur physiology and structure, much as the bat radiation is an aerial exploitation of basic, primitive mammal physiology. Bats are not separated into an independent class merely because they fly. We believe that neither flight nor the species diversity of birds merits separation from dinosaurs on a class level." Think of *Tyrannosaurus,* and thank the old terror as a representative of his group, when you split the wishbone later this month.*

*This article originally appeared in *Natural History,* November, 1977.

27 | Nature's Odd Couples

From Nature's chain whatever link you strike,
Tenth, or ten thousandth, breaks the chain alike.
Alexander Pope,
An Essay on Man (1733)

POPE'S COUPLET EXPRESSES a common, if exaggerated, concept of connection among organisms in an ecosystem. But ecosystems are not so precariously balanced that the extirpation of one species must act like the first domino in that colorful metaphor of the cold war. Indeed, it could not be, for extinction is the common fate of all species—and they cannot all take their ecosystem with them. Species often have as much dependence upon each other as Longfellow's "Ships that pass in the night." New York City might even survive without its dogs (I'm not so sure about the cockroaches, but I'd chance it).

Shorter chains of dependence are more common. Odd couplings between dissimilar organisms form a stock in trade for popularizers of natural history. An alga and a fungus make lichen; photosynthetic microorganisms live in the tissue of reef-building corals. Natural selection is opportunistic; it fashions organisms for their current environments and cannot anticipate the future. One species often evolves an unbreakable dependency upon another species; in an inconstant world, this fruitful tie may seal its fate.

I wrote my doctoral dissertation on the fossil land snails of Bermuda. Along the shores, I would often encounter large hermit crabs incongruously stuffed—big claw protruding—into the small shell of a neritid snail (nerites include the familiar "bleeding tooth"). Why, I wondered, didn't these crabs trade their cramped quarters for more commodious lodgings? After all, hermit crabs are exceeded only by modern executives in their frequency of entry into the real estate market. Then, one day, I saw a hermit crab with proper accommodations—a shell of the "whelk" *Cittarium pica,* a large snail and major food item throughout the West Indies. But the *Cittarium* shell was a fossil, washed out of an ancient sand dune to which it had been carried 120,-000 years before by an ancestor of its current occupant. I watched carefully during the ensuing months. Most hermits had squeezed into nerites, but a few inhabited whelk shells and the shells were always fossils.

I began to put the story together, only to find that I had been scooped in 1907 by Addison E. Verrill, master taxonomist, Yale professor, protégé of Louis Agassiz, and diligent recorder of Bermuda's natural history. Verrill searched the records of Bermudian history for references to living whelks and found that they had been abundant during the first years of human habitation. Captain John Smith, for example, recorded the fate of one crew member during the great famine of 1614–15: "One amongst the rest hid himself in the woods, and lived only on Wilkes and Land Crabs, fat and lusty, many months." Another crew member stated that they made cement for the seams of their vessels by mixing lime from burned whelk shells with turtle oil. Verrill's last record of living *Cittarium* came from kitchen middens of British soldiers stationed on Bermuda during the war of 1812. None, he reported, had been seen in recent times, "nor could I learn that any had been taken within the memory of the oldest inhabitants." No observations during the past seventy years have revised Verrill's conclusion that *Cittarium* is extinct in Bermuda.

As I read Verrill's account, the plight of *Cenobita diogenes* (proper name of the large hermit crab) struck me with that anthropocentric twinge of pain often invested, perhaps improperly, in other creatures. For I realized that nature had condemned *Cenobita* to slow elimination on Bermuda. The neritid shells are too small; only juvenile and very young adult crabs fit inside them—and very badly at that. No other modern snail seems to suit them and a successful adult life requires the discovery and possession (often through conquest) of a most precious and dwindling commodity—a *Cittarium* shell. But *Cittarium*, to borrow the jargon of recent years, has become a "nonrenewable resource" on Bermuda, and crabs are still recycling the shells of previous centuries. These shells are thick and strong, but they cannot resist the waves and rocks forever—and the supply constantly diminishes. A few "new" shells tumble down from the fossil dunes each year—a precious legacy from ancestral crabs that carried them up the hills ages ago—but these cannot meet the demand. *Cenobita* seems destined to fulfill the pessimistic vision of many futuristic films and scenarios: depleted survivors fighting to the death for a last morsel. The scientist who named this large hermit chose well. Diogenes the Cynic lit his lantern and searched the streets of Athens for an honest man; none could he find. *C. diogenes* will perish looking for a decent shell.

This poignant story of *Cenobita* emerged from deep storage in my mind when I heard a strikingly similar tale recently. Crabs and snails forged an evolutionary interdependence in the first story. A more unlikely combination —seeds and dodos—provides the second, but this one has a happy ending.

William Buckland, a leading catastrophist among nineteenth-century geologists, summarized the history of life on a large chart, folded several times to fit in the pages of his popular work *Geology and Mineralogy Considered With Reference to Natural Theology.* The chart depicts victims of mass extinctions grouped by the time of their extirpation. The great animals are crowded together: ichthyosaurs, dinosaurs, ammonites, and pterosaurs in one cluster; mammoths, woolly rhinos, and giant cave bears in another. At the far right, representing modern animals, the dodo stands alone, the first recorded extinction of our era. The dodo, a giant flight-less pigeon (twenty-five pounds or more in weight), lived in fair abundance on the island of Mauritius. Within 200 years of its discovery in the fifteenth century, it had been wiped out—by men who prized its tasty eggs and by the hogs that early sailors had transported to Mauritius. No living dodos have been seen since 1681.

In August, 1977, Stanley A. Temple, a wildlife ecologist at the University of Wisconsin, reported the following re-markable story (but see postscript for a subsequent chal-lenge). He, and others before him, had noted that a large tree, *Calvaria major,* seemed to be near the verge of extinc-tion on Mauritius. In 1973, he could find only thirteen "old, overmature, and dying trees" in the remnant native forests. Experienced Mauritian foresters estimated the trees' ages at more than 300 years. These trees produce well-formed, apparently fertile seeds each year, but none germinate and no young plants are known. Attempts to induce germina-tion in the controlled and favorable climate of a nursery have failed. Yet *Calvaria* was once common on Mauritius; old forestry records indicate that it had been lumbered extensively.

Calvaria's large fruits, about two inches in diameter, con-sist of a seed enclosed in a hard pit nearly half an inch thick. This pit is surrounded by a layer of pulpy, succulent mate-rial covered by a thin outer skin. Temple concluded that *Calvaria* seeds fail to germinate because the thick pit "me-chanically resists the expansion of the embryo within." How, then, did it germinate in previous centuries?

Temple put two facts together. Early explorers reported that the dodo fed on fruits and seeds of large forest trees; in fact, fossil *Calvaria* pits have been found among skeletal remains of the dodo. The dodo had a strong gizzard filled with large stones that could crush tough bits of food. Se-condly, the age of surviving *Calvaria* trees matches the de-mise of the dodo. None has sprouted since the dodo disap-peared almost 300 years ago.

Temple therefore argues that *Calvaria* evolved its unusu-ally thick pit as an adaptation to resist destruction by crushing in a dodo's gizzard. But, in so doing, they became dependent upon dodos for their own reproduction. Tit for tat. A pit thick enough to survive in a dodo's gizzard is a pit too thick for an embryo to burst by its own resources. Thus, the gizzard that once threatened the seed had become its necessary accomplice. The thick pit must be abraded and scratched before it can germinate.

Several small animals eat the fruit of *Calvaria* today, but they merely nibble away the succulent middle and leave the internal pit untouched. The dodo was big enough to swallow the fruit whole. After consuming the middle, dodos would have abraded the pit in their gizzards before regurgitating it or passing it in their feces. Temple cites many analogous cases of greatly increased germination rates for seeds after passage through the digestive tracts of various animals.

Temple then tried to estimate the crushing force of a dodo's gizzard by making a plot of body weight versus force generated by the gizzard in several modern birds. Extrapolating the curve up to a dodo's size, he estimates that *Calvaria* pits were thick enough to resist crushing; in fact, the thickest pits could not be crushed until they had been reduced nearly 30 percent by abrasion. Dodos might well have regurgitated the pits or passed them along before subjecting them to such an extended treatment. Temple took turkeys—the closest modern analogue to dodos—and force-fed them *Calvaria* pits, one at a time. Seven of seventeen pits were crushed by the turkey's gizzard, but the other ten were regurgitated or passed in feces after considerable abrasion. Temple planted these seeds and three of them germinated. He writes: "These may well have been the first *Calvaria* seeds to germinate in more than 300 years." *Calvaria* can probably be saved from the brink of extinction by the propagation of artificially abraded seeds. For once, an astute observation, combined with imaginative thought and experiment, may lead to preservation rather than destruction.

I wrote this essay to begin the fifth year of my regular column in Natural History magazine. I said to myself at the beginning that I would depart from a long tradition of popular writing in natural history. I would not tell the fascinating tales of nature merely for their own sake. I would tie any particular story to a general principle of evolutionary theory: pandas and sea turtles to imperfection as the proof of evolution, magnetic bacteria to principles of scaling, mites that eat their mother from inside to Fisher's theory of sex ratio. But this column has no message beyond the evident homily that things are connected to other things in our complex world—and that local disruptions have wider consequences. I have only recounted these two, related stories because they touched me—one bitterly, the other with sweetness.

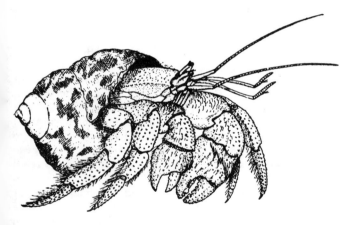

Coenobita diogenes in the shell of *Cittarium*. Drawn from life by A. Verrill in 1900.

Postscript

Some stories in natural history are too beautiful and complex to win general acceptance. Temple's report received immediate publicity in the popular press (*New York Times* and other major newspapers, followed two months later by my article). A year later (March 30, 1979), Dr. Owadally of the Mauritian Forestry Service raised some important doubts in a technical comment published in the professional journal *Science* (where Temple's original article had appeared). I reproduce below, verbatim, both Owadally's comment and Temple's response:

I do not dispute that coevolution between plant and animal exists and that the germination of some seeds may be assisted by their passing through the gut of animals. However, that "mutualism" of the famous dodo and *Calvaria major* (tambalacoque) is an example (*1*) of coevolution is untenable for the following reasons.

1) *Calvaria major* grows in the upland rain forest of Mauritius with a rainfall of 2500 to 3800 mm per annum. The dodo according to Dutch sources roamed over the northern plains and the eastern hills in the Grand Port area—that is, in a drier forest—where the Dutch established their first settlement. Thus it is highly improbable that the dodo and the tambalacoque occurred in the same ecological niche. Indeed, extensive excavations in the uplands for reservoirs, drainage canals, and the like have failed to reveal any dodo remains.

2) Some writers have mentioned the small woody seeds found in Mare aux Songes and the possibility that their germination was assisted by the dodo or other birds. But we now know that these seeds are not tambalacoque but belong to another species of lowland tree recently identified as *Sideroxylon longifolium*.

3) The Forestry Service has for some years been studying and effecting the germination of tambalacoque seeds without avian intervention (*2*). The germination rate is low but not more so than that of many other indigenous species which have, of recent decades, showed a marked deterioration in reproduc-

tion. This deterioration is due to various factors too complex to be discussed in this comment. The main factors have been the depredations caused by monkeys and the invasion by exotic plants.

4) A survey of the climax rain forest of the uplands made in 1941 by Vaughan and Wiehe (3) showed that there was quite a significant population of young tambalacoque plants certainly less than 75 to 100 years old. The dodo became extinct around 1675!

5) The manner in which the tambalacoque seed germinates was described by Hill (4), who demonstrated how the embryo is able to emerge from the hard woody endocarp. This is effected by the swollen embryo breaking off the bottom half of the seed along a well-defined fracture zone.

It is necessary to dispel the tambalacoque-dodo "myth" and recognize the efforts of the Forestry Service of Mauritius to propagate this magnificent tree of the upland plateau.

A. W. OWADALLY

Forestry Service, Curepipe, Mauritius

References and Notes

1. S. Temple, *Science* **197**, 885 (1977).
2. Young *Calvaria major* plants that are 9 months old or more can be seen at the Forest Nursery in Curepipe.
3. R. E. Vaughan and P. O. Wiehe, *J. Ecol.* **19**, 127 (1941).
4. A. W. Hill, *Ann. Bot.* **5**, 587 (1941).

28 March 1978

The plant-animal mutualism that may have existed between the dodo and *Calvaria major* became impossible to prove experimentally after the dodo's extinction. What I pointed out (1) was the possibility that such a relation may have occurred, thus providing an explanation for the extraordinarily poor germination rate in *Calvaria*. I acknowledge the potential for error in historical reconstructions.

I disagree, however, with the conclusion of Owadally

(2) that the dodo and *Calvaria* were geographically separated. There have been virtually no bones of dodos or any other animals found in the uplands of Mauritius not because the animals were never there, but because the island's topography does not cause alluvial deposits there. Catchment basins in certain lowland areas accumulated many bones of animals that were washed into these areas from the surrounding uplands. Accounts of early explorers, summarized by Hachisuka *(3,* p. 85), definitely refer to dodos occurring in the uplands, and Hachisuka makes a point of clarifying the misconception that dodos were strictly coastal birds. Early forestry records from Mauritius *(4)* indicate that *Calvaria* was found in the lowlands as well as on the upland plateau. Although native forests only occur in the uplands today, one of the surviving *Calvaria* trees is located at an elevation of only 150 m. Thus, the dodo and *Calvaria* may have been sympatric, making a mutualistic relation possible.

Taxonomic authorities on sapotaceous plants of the Indian Ocean region recognize seeds of *Calvaria major,* as well as the smaller seeds of *Sideroxylon longifolium,* from alluvial deposits of the Mare aux Songes marsh *(5),* but this has little relevance to the question of mutualism. Mutualistic species will not necessarily be fossilized together.

The Mauritius Forestry Service has only recently succeeded in propagating *Calvaria* seeds, and the unmentioned reason for their recent success strengthens the case for mutualism. Success was achieved when the seeds were mechanically abraded before planting *(6)*. A dodo's digestive tract merely abraded the endocarp naturally the same way the staff of the Mauritius Forestry Service does artificially before the seeds are planted.

The reference Owadally cites *(7)* is equivocal about the age of the surviving *Calvaria* trees because there is no easy way to accurately date them. Coincidently, Wiehe, the coauthor of the paper Owadally cites, was also my source of the estimated age of over 300 years for the surviving trees. I agree that there were more trees surviving in the 1930's than today, which further suppports the notion that *Calvaria major* is a declining species and may have been so since 1681.

I erred in not citing Hill *(8)*. However, Hill does not describe how and under what conditions he induced a seed to germinate. Without these details, his description is of little relevance to the question of mutualism.

STANLEY A. TEMPLE

Department of Wildlife Ecology,
University of Wisconsin–Madison,
Madison 53706

References and Notes

1. S.A. Temple, *Science* **197**, 885 (1977).
2. A. W. Owadally, *ibid.* **203**, 1363 (1979).
3. M. Hachisuka, *The Dodo and Kindred Birds* (Witherby, London, 1953).
4. N. R. Brouard, *A History of the Woods and Forests of Mauritius* (Government Printer, Mauritius, 1963).
5. F. Friedmann, personal communication.
6. A. M. Gardner, personal communication.
7. R. E. Vaughan and P. O. Wiehe, *J. Ecol.* **19**, 127 (1941).
8. A. W. Hill, *Ann. Bot.* **5**, 587 (1941).

I think that Temple has responded adequately (even triumphantly) to Owadally's first three points. As a paleontologist, I can certainly affirm his arguments about the rarity of upland fossils. Our fossil record of upland faunas is exceedingly spotty; the specimens we do possess are generally found in lowland deposits, well worn and washed in from higher ground. Owadally was certainly remiss in not mentioning (point 3) that the Forestry Service abrades its *Calvaria* seeds before they germinate; for the necessity of abrasion lies at the heart of Temple's hypothesis. But Temple was equally remiss in not citing the local Mauritian efforts, which, apparently, predate his own discovery.

Owadally's fourth point, however, represents the potential disproof of Temple's claim. If "quite a significant population" of *Calvaria* trees were less than 100 years old in 1941, then dodos cannot have assisted their germination. Temple denies that so young an age has been demonstrated, and I certainly have no additional insight that can resolve this crucial question.

This exchange highlights a disturbing issue in the transmission of news about science to the public. Many sources cited Temple's original story. I did not find a single mention of the subsequent doubts. Most "good" stories turn out to be false, or at least overextended, but debunking doesn't match the fascination of a clever hypothesis. Most of the "classic" stories of natural history are wrong, but nothing is so resistant to expurgation as textbook dogma.

The debate between Owadally and Temple is too close to call at the moment. I'm rooting for Temple, but if Owadally's fourth point is correct, then the dodo hypothesis will become, in Thomas Henry Huxley's inimitable words, "a beautiful theory, killed by a nasty, ugly little fact."

28 | Sticking Up for Marsupials

I AM ANNOYED that the rapacious ways of my own species have irrevocably prevented me from seeing the dodo in action, for a pigeon as large as a turkey must have been something else, and stuffed, moldy specimens just don't carry conviction. We who revel in nature's diversity and feel instructed by every animal tend to brand *Homo sapiens* as the greatest catastrophe since the Cretaceous extinction. Yet I would argue that the rise of the Isthmus of Panama a mere two to three million years ago must rank as the most devastating biological tragedy of recent times.

South America had been an island continent throughout the Tertiary period (for seventy million years before the onset of continental glaciation). Like Australia, it housed a unique suite of mammals. But Australia was a backwater compared with the range and variety of South American forms. Many survived the onslaught of North American species after the isthmus rose. Some spread and prospered: the opossum moved as far as Canada; the armadillo is still making its way north.

Despite the success of a few, extirpation of the most dramatically different South American forms must be ranked as the dominant effect of contact between mammals of the two continents. Two entire orders perished (we group all modern mammals into about twenty-five orders). Think how our zoos would have been enriched with a liberal sprinkling of notoungulates, a large and diverse group of plant-eating mammals, ranging from rhino-sized *Toxodon,* first exhumed by Charles Darwin on shore leave from the *Beagle,* to rabbit and rodent analogues among the typotheres and hegetotheres. Consider the litopterns with their two subgroups—the large, long-necked camel-like macrauchenids and the most remarkable group of all, the horselike proterotheres. (Proterotheres even repeated some of the evolutionary trends followed by true horses: three-toed *Diadiaphorus* preceded *Thoatherium,* a single-toed species that outdid Man 'O War by reducing its vestigial side toes to a degree never matched by modern horses.) They are all gone forever, victims in large part of faunal disruptions set in motion by the rising isthmus. (Several notoungulates and litopterns survived well into the glacial epoch. They may even have received their *coup de grâce* from early human hunters. Still, I do not doubt that many would still be with us if South America had remained an island.)

The native predators of these South American herbivores also disappeared completely. The modern carnivores of South America, the jaguars and their allies, are all North American interlopers. The indigenous carnivores, believe it or not, were all marsupials (although some flesh-eating niches were occupied by the phororhacids, a remarkable group of giant birds, now also extinct). The marsupial carnivores, although not as diverse as placental carnivores in northern continents, formed an impresive array, from fairly small animals to bear-sized species. One lineage evolved in uncanny parallel with the saber-toothed cats of North America. The marsupial *Thylacosmilus* developed long, stabbing upper canines and a protecting flange of bone on the lower jaw—just like *Smilodon* of the La Brea tar pits.

Although it is not commonly bruited about, marsupials are not doing badly in South America today. North America may only boast the so-called Virginia opossum (actually a South American migrant), but opossums in South America are a rich and varied group of some sixty-five species. In addition, the caenolestids, pouchless "opossum rats," form a separate group with no close affinity to true opossums.

But the third great group of South American marsupials, the carnivorous borhyaenids, were completely wiped out and replaced by northern cats.

The traditional view—though I dedicate this essay to opposing it—attributes the extirpation of carnivorous marsupials to the general inferiority of pouched versus placental mammals. (All living mammals except marsupials and the egg-laying platypus and echidna are placentals.) The argument seems hard to beat. Marsupials flourished only on the isolated island continents of Australia and South America where large placental carnivores never gained a foothold. The early Tertiary marsupials of North America soon disappeared as placentals diversified; South American marsupials took a beating when the Central American corridor opened for placental immigration.

These arguments of biogeography and geological history gain apparent support from the conventional idea that marsupials are anatomically and physiologically inferior to placentals. The very terms of our taxonomy reinforce this prejudice. All mammals are divided into three parts: the egg-laying monotremes are called Prototheria, or premammals; placentals win the prize as Eutheria, or true mammals; the poor marsupials lie in limbo as Metatheria, or middle mammals—not all quite there.

The argument for structural inferiority rests largely upon differing modes of reproduction in marsupials versus placentals, bolstered by the usual smug assumption that different from us is worse. Placentals, as we know and experience, develop as embryos in intimate connection with a mother's body and blood supply. With some exceptions, they are born as reasonably complete and capable creatures. Marsupial fetuses never developed the essential trick that permits extensive development within a mother's body. Our bodies have an uncanny ability to recognize and reject foreign tissues, an essential protection against disease, but a currently intractable barrier to medical procedures ranging from skin grafts to heart transplants. Despite all the homilies about mother love, and the presence of 50 percent maternal genes in offspring, an embryo is still foreign tissue. The maternal immune system must be masked to prevent rejection. Placental fetuses have "learned" to do this; marsupials have not.

Marsupial gestation is very short—twelve to thirteen days in the common opposum, followed by sixty to seventy days of further development in the external pouch. Moreover, internal development does not proceed in intimate connection with the mother, but shielded from her. Two-thirds of gestation occurs within the "shell membrane," a maternal organ that prevents the incursion of lymphocytes, the "soldiers" of the immune system. A few days of placental contact follow, usually via the yolk sac. During this time, the mother mobilizes her immune system, and the embryo is born (or, more accurately, expelled) soon after.

The marsupial neonate is a tiny creature, equivalent in development to a rather early placental embryo. Its head and forelimbs are precociously developed, but the hind limbs are often little more than undifferentiated buds. It must then undertake a hazardous journey, slowly pulling itself along through the relatively great distance to mother's nipples and pouch (we can now understand the necessity of well-developed forelimbs). Our embryonic life within a placental womb sounds altogether easier and unconditionally better.

What challenge can then be offered to these biogeographical and structural accounts of marsupial inferiority? My colleague John A. W. Kirsch has recently marshaled the arguments. Citing work of P. Parker, Kirsch contends that marsupial reproduction follows a different adaptive mode, not an inferior path. True, marsupials never evolved a mechanism to turn off the maternal immune system and permit a completed development within the womb. But early birth may be an equally adaptive strategy. Maternal rejection need not represent a failure of design or lost evolutionary opportunity; it may reflect an ancient and perfectly adequate approach to the rigors of survival. Parker's argument goes right back to Darwin's central contention that individuals struggle to maximize their own reproductive success, that is, to increase the representation of their own genes in future generations. Several highly divergent, but equally successful, strategies can be followed in (unconscious) pursuit of this goal. Placentals invest a great deal of time and energy in offspring before their birth. This commitment does increase the chance of an offspring's success, but the placental mother also takes a risk: if she should lose her litter, she has irrevocably expended a large portion of her life's reproductive effort for no evolutionary gain. The marsupial mother pays a much higher toll in neonatal death, but her reproductive cost is small. Gestation has been very short and she may breed again in the same season. Moreover, the tiny neonate has not placed a great drain upon her energetic resources, and has subjected her to little danger in a quick and easy birth.

Turning to biogeography, Kirsch challenges the usual assumption that Australia and South America were refugia for inferior beasts that couldn't hang on in the placental world of the Northern Hemisphere. He views their southern diversity as a reflection of success in their ancestral homeland, not as a feeble effort in peripheral territory. His argument relies upon M. A. Archer's claim for close genealogical relationship between borhyaenids (South American marsupial carnivores) and thylacines (marsupial carnivores of the Australian region). Taxonomists have previously regarded these two groups as an example of evolutionary convergence—separate development of similar adaptations (as in the marsupial and placental saber-tooths, mentioned previously). In fact, taxonomists have viewed the Australian and South American radiation of marsupials as completely independent events, following the separate invasion of both continents by primitive marsupials pushed out from northern lands. But if borhyaenids and thylacines are closely related, then the southern continents must have exchanged some of their products, probably via Antarctica. (In our new geology of drifting continents, southern hemisphere lands were much closer together when mammals rose to prominence, following the dinosaurs' demise.) A more parsimonious view imagines an Australian center of origin for marsupials and a dispersal to South America following the evolution of thylacinids, rather than two separate marsupial invasions of South America—borhyaenid ancestors from Australia, and all the others from North America. Although the simplest explanations are not always true in our wondrously complex world, Kirsch's arguments do cast considerable doubt on the usual assumption that marsupial homelands are refugia, not centers of origin.

Yet I must confess that this structural and biogeographical defense of marsupials falters badly before one cardinal fact, prominently featured above: the Isthmus of Panama rose, placental carnivores invaded, marsupial carnivores quickly perished, and the placentals took over. Does this not speak for clear competitive superiority of North American placental carnivores? I could sneak around this unpleasant fact by ingenious conjecture, but I prefer to admit it. How then can I continue to defend marsupial equality?

Although the borhyaenids lost big, I find no scrap of evidence to attribute defeat to their status as marsupials. I prefer an ecological argument predicting hard times for any indigenous group of South American carnivores, marsupial or placental. The real victims happened to be marsupials, but this taxonomic fact may be incidental to a fate sealed for other reasons.

R. Bakker has been studying the history of mammalian carnivores throughout the Tertiary. Integrating some new ideas with conventional wisdom, he finds that the northern placental carnivores experienced two kinds of evolutionary "tests." Twice, they suffered short periods of mass extinction, and new groups, perhaps with greater adaptive flexibility, took over. During times of continuity, high diversity of both predators and prey engendered intense competition and strong evolutionary trends for improvement in feeding (quick ingestion and efficient slicing) and locomotion (high acceleration in ambush predators, endurance in long-distance hunters). South American and Australian carnivores were tested in neither way. They suffered no mass extinctions, and the original incumbents persisted. Diversity never approached northern levels, and competition remained less intense. Bakker reports that their levels of morphological specialization for running and feeding lie far below those of northern carnivores living at the same time.

H. J. Jerison's studies of brain size provide an impressive confirmation. On northern continents, placental predators and prey evolved successively larger brains throughout the Tertiary. In South America, both marsupial carnivores and their placental prey quickly plateaued at about 50 percent of brain weight for average modern mammals of the same body sizes. Anatomical status as marsupial or placental seems to make no difference; a relative history of evolutionary challenge may be crucial. If, by happenstance, northern carnivores had been marsupials and southern carnivores placentals, I suspect that the outcome of isthmian exchange would still have been a rout for South America. North American faunas were continually tested in the fiery furnaces of mass destruction and intense competition. The South American carnivores were never strongly challenged. When the Isthmus of Panama rose, they were weighed in the evolutionary balance for the first time. Like Daniel's king, they were found wanting.

8 | Size and Time

29 | Our Allotted Lifetimes

J. P. MORGAN, MEETING with Henry
Ford in E. L. Doctorow's *Ragtime,* praises the assembly line
as a faithful translation of nature's wisdom:

> Has it occurred to you that your assembly line is not
> merely a stroke of industrial genius but a projection of
> organic truth? After all, the interchangeability of parts
> is a rule of nature. . . . All mammals reproduce in the
> same way and share the same designs of self-nourish-
> ment, with digestive and circulatory systems that are
> recognizably the same, and they enjoy the same senses.
> . . . Shared design is what allows taxonomists to classify
> mammals as mammals.

An imperious tycoon should not be met with equivoca-
tion; nonetheless, I can only reply "yes, and no" to Morgan's pronouncement. Morgan was wrong if he thought
that large mammals are geometric replicas of smaller relatives. Elephants have relatively smaller brains and thicker
legs than mice, and these differences record a general rule
of mammalian design, not the idiosyncrasies of particular
animals.

But Morgan was right in arguing that large animals are
essentially similar to small members of their group. The
similarity, however, does not reside in a constant shape.
The basic laws of geometry dictate that animals must

change their shape in order to work the same way at different sizes. Galileo himself established the classic example in 1638: the strength of an animal's leg is a function of its cross-sectional area (length × length); the weight that legs must support varies as the animal's volume (length × length × length). If mammals did not increase the relative thickness of their legs as they got larger, they would soon collapse (since body weight would increase so much faster than the supporting strength of limbs). To remain the same in function, animals must change their form.

The study of these changes in form is called "scaling theory." Scaling theory has uncovered a striking regularity of changing shape over the 25-millionfold range of mammalian weight from shrew to blue whale. If we plot brain weight versus body weight for all mammals on the so-called mouse-to-elephant (or shrew-to-whale) curve, very few species deviate far from a single line expressing the general rule: brain weight increases only two-thirds as fast as body weight as we move from small to large mammals. (We share with bottle-nosed dolphins the honor of greatest upward deviance from the curve.)

We can often predict these regularities from the basic physics of objects. The heart, for example, is a pump. Since all mammalian hearts work in essentially the same way, small hearts must pump considerably faster than large ones (imagine how much faster you could work a finger-sized, toy bellows than the giant model that fuels a blacksmith's forge or an old-fashioned organ). On the mouse-to-elephant curve for mammals, the length of a heartbeat increases between one-fourth and one-third as fast as body weight as we move from small to large mammals. The generality of this conclusion has recently been affirmed in an interesting study by J. E. Carrel and R. D. Heathcote on the scaling of heart rate in spiders. They used a cool laser beam to illuminate the hearts of resting spiders and drew a crab spider-to-tarantula curve for eighteen species spanning nearly a thousandfold range of body weight. Again, scaling is regular with heart rate increasing four-tenths as fast as body weight (.409 times as fast, to be exact).

We may extend this conclusion for hearts to a general statement about the pace of life in small versus large animals. Small animals tick through life far more rapidly than large animals—their hearts work more quickly, they breathe more frequently, their pulse beats much faster. Most importantly, metabolic rate, the so-called fire of life, increases only three-fourths as fast as body weight in mammals. To keep themselves going, large mammals do not need to generate as much heat per unit of body weight as small animals. Tiny shrews move frenetically, eating nearly all their waking lives to keep their metabolic fire burning at the maximal rate among mammals; blue whales glide majestically, their hearts beating the slowest rhythm among active, warmblooded creatures.

The scaling of lifetime among mammals suggests an intriguing synthesis of these disparate data. We have all had enough experience with mammalian pets of various sizes to understand that small mammals tend to live for a shorter time than large ones. In fact, mammalian lifetime scales at about the same rate as heartbeat and breath time—between one-fourth and one-third as fast as body weight as we move from small to large animals. (*Homo sapiens* emerges from this analysis as a very peculiar animal. We live far longer than a mammal of our body size should. In essay 9, I argue that humans evolved by an evolutionary process called "neoteny"—the preservation in adults of shapes and growth rates that characterize juvenile stages of ancestral primates. I also believe that neoteny is responsible for our elevated longevity. Compared with other mammals, all stages of human life arrive "too late." We are born as helpless embryos after a long gestation; we mature late after an extended childhood; we die, if fortune be kind, at ages otherwise reached by warmblooded animals only at the very largest sizes.)

Usually, we pity the pet mouse or gerbil that lived its full span of a year or two at most. How brief its life, while we endure for the better part of a century. As the main theme of this essay, I want to argue that such pity is misplaced (our personal grief, of course, is quite another matter; with this, science does not deal). Morgan was right in *Ragtime*—small and large mammals are essentially similar. Their lifetimes are scaled to their life's pace, and all endure for approximately the same amount of biological time. Small mammals tick fast, burn rapidly, and live for a short time; large mammals live long at a stately pace. Measured by their own internal clocks, mammals of different sizes tend to live for the same amount of time.

We are prevented from grasping this important and comforting concept by a deeply ingrained habit of Western thought. We are trained from earliest memory to regard absolute Newtonian time as the single valid measuring stick in a rational and objective world. We impose our kitchen clock, ticking equably, upon all things. We marvel at the quickness of a mouse, express boredom at the torpor of a hippopotamus. Yet each is living at the appropriate pace of its own biological clock.

I do not wish to deny the importance of absolute, astronomical time to organisms (see essay 31). Animals must measure it to lead successful lives. Deer must know when to regrow their antlers, birds when to migrate. Animals track the day–night cycle with their circadian rhythms; jet lag is the price we pay for moving much faster than nature intended.

But absolute time is not the appropriate measuring stick for all biological phenomena. Consider the magnificent song of the humpback whale. E. O. Wilson has described the awesome effect of these vocalizations: "The notes are eerie yet beautiful to the human ear. Deep basso groans and almost inaudibly high soprano squeaks alternate with repetitive squeals that suddenly rise or fall in pitch." We do not know the function of these songs. Perhaps they enable whales to find each other and to stay together during their annual transoceanic migrations. Perhaps they are the mating songs of courting males.

Each whale has its own characteristic song; the highly complex patterns are repeated over and over again with great faithfulness. No scientific fact that I have learned in the last decade struck me with more force than Roger S.

Payne's report that the length of some songs may extend for more than half an hour. I have never been able to memorize the five-minute first *Kyrie* of the B-minor Mass (and not for want of trying); how could a whale sing for thirty minutes and then repeat itself accurately? Of what possible use is a thirty-minute repeat cycle—far too long for a human to recognize; we would never grasp it as a single song (without Payne's recording machinery and much study after the fact). But then I remembered the whale's metabolic rate, the enormously slow pace of its life compared with ours. What do we know about a whale's perception of thirty minutes? A humpback may scale the world to its own metabolic rate; its half-hour song may be our minute waltz. From any point of view, the song is spectacular; it is the most elaborate single display so far discovered in any animal. I merely urge the whale's point of view as an appropriate perspective.

We can provide some numerical precision to support the claim that all mammals, on average, live for the same amount of biological time. In a method developed by W. R. Stahl, B. Günther, and E. Guerra in the late 1950s and early 1960s, we search the mouse-to-elephant equations for biological properties that scale at the same rate against body weight. For example, Günther and Guerra give the following equations for mammalian breath time and heartbeat time versus body weight.

$$\text{breath time} = .0000470 \text{ body}^{0.28}$$
$$\text{heartbeat time} = .0000119 \text{ body}^{0.28}$$

(Nonmathematical readers need not be overwhelmed by the formalism. The equations simply state that both breath time and heartbeat time increase about .28 times as fast as body weight as we move from small to large mammals.) If we divide the two equations, body weight cancels out because it is raised to the same power in both.

$$\frac{\text{breath time}}{\text{heartbeat time}} = \frac{.0000470 \text{ body}^{0.28}}{.0000119 \text{ body}^{0.28}} = 4.0$$

This states that the ratio of breath time to heartbeat time is 4.0 in mammals of any body size. In other words, all mammals, whatever their size, breathe once for each four heart-beats. Small mammals breathe and beat their hearts faster than large mammals, but both breath and heart slow up at the same relative rate as mammals get larger.

Lifetime also scales at the same rate as body weight (.28 times as fast as we move from small to large mammals). This means that the ratio of both breath time and heartbeat time to lifetime is also constant over the entire range of mammalian size. When we perform a calculation similar to the one above, we find that all mammals, regardless of their size, tend to breathe about 200 million times during their lives (their hearts, therefore, beat about 800 million times). Small mammals breathe fast, but live for a short time. Measured by the internal clocks of their own hearts or the rhythm of their own breathing, all mammals live the same time. (Astute readers, after counting their breaths or taking their pulses, may have calculated that they should have died long ago. But *Homo sapiens* is a markedly deviant mammal in more ways than braininess alone. We live about three times as long as mammals of our body size "should," but we breathe at the "right" rate and thus live to breathe about three times as often as an average mammal of our body size. I regard this excess of living as a happy consequence of neoteny.)

The mayfly lives but a day as an adult. It may, for all I know, experience that day as we live a lifetime. Yet all is not relative in our world, and such a short glimpse of it guarantees distortion in interpreting events ticking on longer scales. In a brilliant metaphor, the pre-Darwinian evolutionist Robert Chambers wrote in 1844 of a mayfly watching the metamorphosis of a tadpole into a frog:

> Suppose that an ephemeron [a mayfly], hovering over a pool for its one April day of life, were capable of observing the fry of the frog in the waters below. In its aged afternoon, having seen no change upon them for such a long time, it would be little qualified to conceive that the external branchiae [gills] of these creatures were to decay, and be replaced by internal lungs, that

feet were to be developed, the tail erased, and the animal then to become a denizen of the land.

Human consciousness arose but a minute before midnight on the geologic clock. Yet we mayflies try to bend an ancient world to our purposes, ignorant perhaps of the messages buried in its long history. Let us hope that we are still in the early morning of our April day.

30 | Natural Attraction: Bacteria, the Birds and the Bees

THE FAMOUS WORDS "blessed art thou among women" were uttered by the angel Gabriel as he announced to Mary that she would conceive by the Holy Spirit. In medieval and Renaissance painting, Gabriel bears the wings of a bird, often elaborately spread and adorned. While visiting Florence last year, I became fascinated by the "comparative anatomy" of Gabriel's wings as depicted by the great painters of Italy. The faces of Mary and Gabriel are so beautiful, their gestures often so expressive. Yet the wings, as painted by Fra Angelico or by Martini, seem stiff and lifeless, despite the beauty of their intricate feathering.

But then I saw Leonardo's version. Gabriel's wings are so supple and graceful that I scarcely cared to study his face or note the impact he had upon Mary. And then I recognized the source of the difference. Leonardo, who studied birds and understood the aerodynamics of wings, had painted a working machine on Gabriel's back. His wings are both beautiful and efficient. They have not only the right orientation and camber, but the correct arrangement of feathers as well. Had he been just a bit lighter, Gabriel might have flown without divine guidance. In contrast, the other Gabriels bear flimsy and awkward ornaments that could never work. I was reminded that aesthetic and functional beauty often go hand in hand (or rather arm in arm in this case).

In the standard examples of nature's beauty—the cheetah running, the gazelle escaping, the eagle soaring, the tuna coursing, and even the snake slithering or the inchworm inching—what we perceive as graceful form also represents an excellent solution to a problem in physics. When we wish to illustrate the concept of adaptation in evolutionary biology, we often try to show that organisms unconsciously "know" physics—that they have evolved remarkably efficient machines for eating and moving. When Mary asked Gabriel how she could possibly conceive, "seeing I know not a man," the angel replied: "For with God nothing shall be impossible." Many things are impossible for nature. But what nature can do, she often does surpassingly well. Good design is usually expressed by correspondence between an organism's form and an engineer's blueprint.

I recently encountered an even more striking example of good design: an organism that builds an exquisite machine directly within its own body. The machine is a magnet; the organism, a "lowly" bacterium. When Gabriel departed, Mary went to visit Elizabeth, who had also conceived with a bit of help from on high. Elizabeth's babe (the future John the Baptist) "leaped in her womb" and Mary pronounced the *Magnificat*, including the line (later set so incomparably by Bach) *et exaltavit humilis*—"and he hath exalted them of low degree." The tiny bacteria, simplest in structure among organisms, inhabitants of the first rung on traditional (and fallacious) ladders of life, illustrate in a few microns all the wonder and beauty that some organisms require meters to express.

In 1975, University of New Hampshire microbiologist Richard P. Blakemore discovered "magnetotactic" bacteria in sediments near Woods Hole, Massachusetts. (Just as geotactic organisms orient toward gravitational fields and phototactic creatures toward light, magnetotactic bacteria align themselves and swim in preferred directions within magnetic fields.) Blakemore then spent a year at the University of Illinois with microbiologist Ralph Wolfe and managed to isolate and culture a pure strain of magnetotactic bacteria. Blakemore and Wolfe then turned to an expert on the physics of magnetism, Richard B. Frankel of the National Magnet Laboratory at M.I.T. (I thank Dr. Frankel for his patient and lucid explanation of their work.)

Frankel and his colleagues found that each bacterium builds within its body a magnet made of twenty or so opaque, roughly cubic particles, measuring about 500 angstroms on a side (an angstrom is one ten-millionth of a millimeter). These particles are made primarily of the magnetic material Fe_3O_4, called magnetite, or lodestone. Frankel then calculated the total magnetic moment per bacterium and found that each contained enough magnetite to orient itself in the earth's magnetic field against the disturbing influence of Brownian motion. (Particles small enough to be unaffected by the gravitational fields that stabilize us

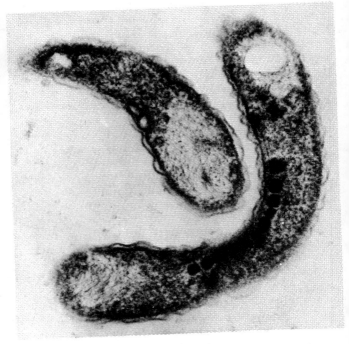

A magnetotactic bacterium with its chain of tiny magnets (X 40,000)   D. L. BALKWILL AND D. MARATEA

or by the surface forces that affect objects of intermediate size are buffeted in a random manner by thermal energy of the medium in which they lie suspended. The "play" of dust particles in sunlight provides a standard illustration of Brownian motion.)

The magnetotactic bacteria have built a remarkable machine, using virtually the only configuration that could work as a compass within their tiny bodies. Frankel explains why the magnetite must be arranged as particles and why the particles must be about 500 angstroms on a side. To work as an efficient compass, magnetite must be present as so-called single domain particles, that is, as bits with a single magnetic moment, containing opposite north- and south-seeking ends. The bacteria contain a chain of such particles, oriented with their magnetic moments north pole to the next south pole along the row—"like the elephants head to tail in a circus finale," as Frankel states. In this way, the entire chain of particles operates as a single magnetic dipole with north- and south-seeking ends.

If the particles were a bit smaller (less than 400 angstroms on a side), they would be "superparamagnetic"—a big word indicating that thermal energy at room temperature would cause internal reorientation of the particle's magnetic moment. On the other hand, if particles were greater than 1,000 angstroms on a side, separate magnetic domains pointing in different directions would form *within* the particle. This "competition" would reduce or cancel the particle's overall magnetic moment. Thus, Frankel concludes, "the bacteria have solved an interesting problem in physics by producing particles of magnetite of just the right size for a compass, of dimension 500 angstroms."

But evolutionary biology is preeminently the science of "why," and we must ask what such a small creature could possibly do with a magnet. Since a bacterium's cruising range is probably a few inches for the few minutes of its existence, I find it hard to believe that oriented motion in a north or south direction can play any role in its repertoire of adaptive traits. But what preferred direction of motion might make a difference? Frankel suggests, quite plausibly in my view, that an ability to move *down* might be crucial for such a bacterium—for down is the direction of sediments in aquatic environments, and down might lead to a region of preferred oxygen pressure. In this instance, "them of low degree" might wish to debase themselves even further.

But how does a bacterium know which way is down? With the smug prejudices of our enormous selves, we might think the question inane for its obvious answer: all they have to do is stop whatever they are doing and fall. Not at all. We fall because gravity affects us. Gravity—the standard example of a "weak force" in physics—influences us only because we are large. We live in a world of competing forces, and the relative strength of these forces depends primarily upon the size of objects affected by them. For familiar creatures of macroscopic dimensions, the ratio of surface area to volume is crucial. This ratio decreases continually as an organism grows, since areas increase as length squared and volumes as length cubed. Small creatures, insects for example, live in a world dominated by forces acting on their surfaces. Some can walk on water or hang upside down from a ceiling because surface tension is so strong and the gravitational force that might pull them down so weak. Gravitation works on volumes (or, to be more precise, upon masses that are proportional to volumes in a constant gravitational field). Gravitation rules us with our low ratio of surface to volume. But it troubles an insect very little—and a bacterium not at all.

The world of a bacterium is so unlike our own that we must abandon all our certainties about the way things are and start from scratch. Next time you see *Fantastic Voyage* on the tube, take your eyes off Raquel Welch and the predaceous white blood corpuscle long enough to ponder how the miniaturized adventurers would really fare as microscopic objects within a human body (they behave just like regular folks in the film). They would, first of all, be subject to shocks of the Brownian motion, thus making the film something of a random blur. Also, as Isaac Asimov pointed out to me, their ship could not run on its propeller, since blood is too viscous at such a scale. It should have, he said, a flagellum—like a bacterium.

D'Arcy Thompson, premier student of scaling since Galileo, urged us to set aside our prejudices if we would understand the world of a bacterium. In his masterpiece, *Growth and Form* (published in 1942 but still in print), he ends his chapter "On Magnitude" in his incomparable prose:

> Life has a range of magnitude narrow indeed compared to that with which physical science deals; but it is wide enough to include three such discrepant conditions as those in which a man, an insect and a bacillus have their being and play their several roles. Man is ruled by gravitation, and rests on mother earth. A water-beetle finds the surface of a pool a matter of life and death, a perilous entanglement or an indispensable support. In a third world, where the bacillus lives, gravitation is forgotten, and the viscosity of the liquid, the resistance defined by Stokes's law, the molecular shocks of the Brownian movement, doubtless also the electric charges of the ionized medium, make up the physical environment and have their potent and immediate influence upon the organism. The predominant factors are no longer those of our scale; we have come to the edge of a world of which we have no experience, and where all our preconceptions must be recast.

So how does a bacterium know which way is down? We use magnets for horizontal orientation so exclusively that we often forget (in fact, I suspect many of us do not know) that the earth's magnetic field also has a vertical component, its strength depending upon latitude. (We damp out the vertical deflection in building compasses because it doesn't interest us. As large creatures ruled by gravitation, we know which way is down. Only at our scale could folly be personified as not knowing "which way is up.") A compass needle follows the earth's lines of force. At the equator, these lines are horizontal to the surface. Toward the poles, they dip more and more strongly *into* the earth. At the magnetic pole itself, the needle points straight down. At my latitude in Boston, the vertical component is actually stronger than the horizontal. A bacterium, swimming north as a free compass needle, also swims down at Woods Hole.

This putative function for a bacterial compass is pure speculation at the moment. But if these bacteria use their magnets primarily to swim down (rather than to find each other, or to do Lord knows what, if anything, in their unfamiliar world), then we can make some testable predictions. Members of the same species, living in natural populations adapted to life at the equator, will probably not make magnets, for here a compass needle has no vertical component. In the Southern Hemisphere, magnetotactic bacteria should display reversed polarity and swim in the direction of their south-seeking pole.

Magnetite has also been reported as a component of several larger organisms, all of which perform remarkable feats of horizontal orientation—the conventional use of a compass for familiar creatures of our scale. Chitons, eight-plated relatives of clams and snails, live primarily on rocks near sea level in tropical regions. They scrape food from the rocks with a long file called a radula—and the tips of the radular teeth are made of magnetite. Many chitons make substantial excursions from a living site, but "home" back to the precise spot thereafter. The idea that they might use their magnetite as an orienting compass suggests itself, but the evidence so far offers no support. It is not even clear that chitons have enough magnetite to perceive the earth's field, and Frankel tells me that their particles are mostly above the single domain limit.

Some bees have magnetite in their abdomens, and we know that they are affected by the earth's magnetic field (see article by J. L. Gould, no relation, J. L. Kirschvink, and K. S. Defeyes in bibliography). Bees do their famous dance on the vertical surface of their honeycomb by converting the orientation of their flight to food in relation to the sun into an angle danced with respect to gravity. If the comb is turned so that bees must dance on a horizontal surface, where they cannot express direction in gravitational terms, they become disoriented at first. Finally, after several weeks, they align their dances to the magnetic compass. Moreover, a swarm of bees, placed into an empty hive without cues for orientation, build their comb in the magnetic direction it occupied in their parental hive. Pigeons, certainly no duffers at homing, build a structure made of magnetite between their brain and skull. This magnetite exists as single domains and can therefore function as a magnet (see C. Walcott *et al.* in bibliography).

The world is full of signals that we don't perceive. Tiny creatures live in a different world of unfamiliar forces. Many animals of our scale greatly exceed our range of perception for sensations familiar to us. Bats avoid obstacles by bouncing sound off them at frequencies that I cannot hear, although some people can. Many insects see into the ultraviolet and follow the "invisible" nectar guides of flowers to sources of food for them and pollen that they will carry to the next flower for fertilization (plants build these orienting color streaks for their own advantages, not to convenience the insects).

What an imperceptive lot we are. Surrounded by so much, so fascinating and so real, that we do not see (hear, smell, touch, taste) in nature, yet so gullible and so seduced by claims for novel power that we mistake the tricks of mediocre magicians for glimpses of a psychic world beyond our ken. The paranormal may be a fantasy; it is certainly a haven for charlatans. But "parahuman" powers of perception lie all about us in birds, bees, and bacteria. And we can use the instruments of science to sense and understand what we cannot directly perceive.

Postscript

In asking why bacteria might build magnets within their bodies, Frankel speculated cogently that swimming north could make little difference to such a tiny creature, but that swimming *down* (another consequence of life around a compass at mid to high latitudes in the northern hemisphere) could be very important indeed. This led me to predict that if Frankel's explanation be valid, magnetic bacteria in the Southern Hemisphere should swim *south* in order to swim down—that is, their polarity should be reversed relative to northern hemisphere relatives.

In March, 1980, Frankel sent me a preprint of a paper with colleagues R.P. Blakemore and A.J. Kalmijn. They travelled to New Zealand and Tasmania in order to test the magnetic polarity of southern hemisphere magnetic bacteria. Indeed, they all swam south and down—an impressive confirmation of Frankel's hypothesis and the basis of my essay.

They also performed an interesting experiment, providing further confirmation of another kind. They collected magnetic bacteria at Woods Hole, Massachusetts, divided the sample of north-swimming cells into two parts. They cultured one subsample for several generations in a chamber of normal polarity, but grew the other in a chamber of reversed polarity to simulate Southern Hemisphere conditions. Sure enough, after several weeks, north-swimming cells continued to predominate in the chamber of normal polarity. But in the chamber with reversed polarity, south-swimming cells now formed a majority. Since bacterial cells do not change polarity during their lifetimes, this dramatic change is probably the result of strong natural selection for the ability to swim down. Presumably, both north and south-swimming cells originate in each chamber, but selection quickly weeds out individuals that cannot swim down.

Frankel tells me that he is now off to the geomagnetic equator to see what happens where the magnetic field has no downward component at all.

31 | Time's Vastness

2:00 A.M., Jan. 1, 1979

I WILL NEVER forget Toscanini's last concert—the night that the greatest maestro of them all, the man who held all Western music in his infallible memory, faltered for a few seconds and lost his place. If heroes were truly invulnerable, how could they compel our interest? Siegfried must have a mortal shoulder, Achilles a heel, Superman kryptonite.

Karl Marx remarked that all historical events occur twice, first as tragedy, the second time as farce. If Toscanini's lapse was tragic (in the heroic sense), then I witnessed the farce just two hours ago. I listened while the ghost of Guy Lombardo missed a beat. For the first time in God only knows how many years, that smooth sound, that comfortable welcome to the New Year, fell apart for a mysterious moment. As I learned later, someone forgot to tell Guy about the special 61-second minute that ended 1978; he started too early and could not compensate with unnoticed grace.

This second, added for internal bookkeeping to synchronize atomic and astronomical clocks, received wide press coverage, virtually all of it in a jocular vein. And why not—good news is rare enough these days. Most reports pushed the same theme: they twitted scientists about their concern for consummate accuracy. After all, how can so trifling a span of time as a single second matter?

I then remembered another figure, 1/50,000 second per year. This figure, an ant before the behemoth of a full second, is the annual rate of deceleration in the earth's rotation due to tidal friction. I will attempt to show just how important such an "insignificant" number can be in the fullness of geological time.

We have known for a long time that the earth is slowing down. Edmund Halley, godfather to the famous comet and Astronomer Royal of England early in the eighteenth century, noted a systematic discrepancy between the recorded position of ancient eclipses and their predicted areas of visibility based on the earth's rate of rotation in his time. He calculated that this disparity could be resolved by assuming a faster rotation in the past. Halley's calculations have been refined and reanalyzed many times, and eclipse records suggest an approximate rate of two milliseconds per century for rotational slowing during the past few thousand years.

Halley proposed no adequate reason for this deceleration. Immanuel Kant, a versatile fellow indeed, supplied the correct explanation later in the eighteenth century. Kant implicated the moon and argued that tidal friction had slowed the earth down. The moon pulls the waters of the earth toward it in a tidal bulge. This bulge remains oriented toward the moon as the earth rotates under it. From our point of view as earthbound observers, high tide moves steadily westward around the earth. This tide, moving continuously across land and sea (for continents have their minor tides as well), creates a great deal of friction. Astronomers Robert Jastrow and M. H. Thompson write: "A huge quantity of energy is dissipated in this friction each day. If the energy could be recovered for useful purposes, it would be sufficient to supply the electrical power requirements of the entire world several times over. The energy is actually dissipated in the turbulence of coastal waters plus a small degree of heating of the rocks in the crust of the earth."

But tidal friction has another effect, virtually invisible on the scale of our lives, but a major factor in the earth's history. It acts as a brake upon the spinning earth, slowing the earth's rotation at the leisurely rate of about two milliseconds per century, or 1/50,000 second per year.

Braking by tidal friction has two correlated and intriguing effects. First, the number of days in a year should be decreasing through time. The length of a year seems to be essentially constant relative to the official cesium clock. Its invariance is affirmed both empirically, by astronomical measurement, and theoretically. We might predict that a solar tide should slow the earth's revolution just as the lunar tide slows its rotation. But solar tides are quite weak, and the earth, hurtling through space, has such an enormous moment of inertia that the year increases by no more than three seconds per billion years. Here we finally have a figure that we can safely ignore—half a minute from the origin of the earth to its destruction by an exploding sun some five billion years hence!

Second, as the earth loses angular momentum in slowing down, the moon—obedient to the law of conservation of angular momentum for the earth–moon system—must pick up what the earth loses. The moon does this by revolving around the earth at a greater and greater distance. In other words, the moon has been steadily receding from the earth.

If the moon looks big now, low on the horizon on a crisp October night, you should have been around to see what the trilobites saw 550 million years ago. G. H. Darwin, noted astronomer and second son of Charles, first developed this idea of lunar recession. He believed that the moon had been wrenched from the Pacific Ocean, and he extrapolated its present rate of recession back to determine the time of this convulsive birth. (It does fit, but thanks to plate tectonics, we now know that the Pacific is not a permanent hole, but a configuration of the geological moment.)

In short, tidal friction induced by the moon entails two coupled consequences through time: slowing the earth's rotation to decrease the number of days per year, and increasing the distance between earth and moon.

Astronomers have long known about these phenomena in theory; they have also measured them directly over geological microseconds. But until recently, no one has known how to gauge their effects over long stretches of geological time. A simple backward extrapolation of the current rate will not suffice because intensity of braking depends upon the configuration of continents and oceans. The most effective braking occurs when tides sweep across shallow seas; the least effective when tides move with comparatively little friction over deep oceans and land. Shallow seas are not prominent features of our present earth, but they covered millions of square miles at various times in the past. The high tidal friction of those times may be matched by very slow deceleration at other times, particularly when all the continents coalesced into a single Pangaea. The pattern of rotational slowing through time therefore becomes more a geological than an astronomical problem.

I am delighted to report that my own brand of geology has yielded, albeit ambiguously, the required information—for some fossils record in their patterns of growth the astronomical rhythms of ancient times. The haughty and high-riding mathematicians and experimentalists of modern geophysics do not often take a bow toward a lowly fossil. Yet one prominent student of the earth's rotation has written "It appears that paleontology comes to the rescue of the geophysicist."

For more than a hundred years, paleontologists had occasionally noted regularly spaced growth lines on some of their fossils. Some had suggested that they might reflect astronomical periods of days, months, or years—much like tree rings. Yet no one had done anything with these observations. Throughout the 1930s Ting Ying Ma, a somewhat visionary, highly speculative, but infallibly interesting Chinese paleontologist, studied annual bands in fossil corals to determine the position of ancient equators. (Corals living at the equator in regimes of nearly constant temperature should not show the seasonal bands; the higher the latitude, the stronger the bands.) But no one had studied the very fine laminations that often occur by the hundreds per band.

In the early 1960s, Cornell paleontologist John West Wells realized that these very fine striations might record days (slow growth at night versus faster growth during daylight, much as trees produce annual bands of alternating slow winter and rapid summer growth). He studied a modern coral with both coarse (presumably annual) and very fine banding, and he counted an average of about 360 fine lines to each coarse band. He concluded that the fine lines are daily.

Wells then searched his collection for fossil corals sufficiently well preserved to retain all their fine bands. He found very few, but they enabled him to make one of the most interesting and important observations in the history of paleontology: a group of corals about 370 million years old had an average of just under 400 fine lines per coarse band. These corals had witnessed a year of nearly 400 days. Direct, geological evidence had finally been found for an old astronomical theory.

But Wells's corals had affirmed only half the story—increasing length of day. The other half, recession of the moon, required fossils with daily and monthly banding; for if the moon had been much closer in the past, it would have revolved around the earth in a much shorter time than it does today. The ancient lunar month should have contained fewer than the 29.53 solar days of the present month.

Since Wells published his famous paper on "Coral Growth and Geochronometry" in 1963, several claims have been entered for lunar periodicities as well. Most recently, Peter Kahn, a paleontologist from Princeton, and Stephen Pompea, a physicist from Colorado State University, have argued that the key to lunar history lies with one of everybody's favorite creatures, the chambered nautilus. The nautilus shell is divided into regular internal partitions called septa. These same septa, and the beauty of their construction, inspired Oliver Wendell Holmes to exhort us, by analogy, to do better with our internal lives:

Build thee more stately mansions, O my soul,
As the swift seasons roll!
Leave thy low-vaulted past!
Let each new temple, nobler than the last,

Shut thee from heaven with a dome more vast,
Till thou at length art free,
Leaving thine outgrown shell by life's unresting sea!

I am happy to report that nautiloid septa may have extended their utility beyond Holmes's musings on immortality and O'Neill's cribbing of a title for a play. For Kahn and Pompea counted the finer growth lines on the exterior of *Nautilus*'s shell and found that each chamber (the space between successive septa) contains an average of thirty fine lines, with little variation either among shells or on successive chambers of single shells. Since *Nautilus*, living in deep Pacific waters, migrates daily in response to the solar cycle (it moves towards the surface at night), Kahn and Pompea suggest that the fine lines record days. The secretion of septa may be entrained to a lunar cycle. Many animals, including humans of course, have lunar cycles, usually tied to breeding.

Nautiloids are quite common as fossils (the modern chambered nautilus is sole survivor of a very diverse group). Kahn and Pompea counted lines per chamber in twenty-five nautiloids ranging in age from 25 to 420 million years. They argue for a regular decrease in lines per chamber from thirty today, to about twenty-five for the youngest fossils, to only nine or so for the oldest. If the moon circled the earth in only nine solar days 420 million years ago (when the day only contained twenty-one hours), then it must have been much closer. Cranking through some equations, Kahn and Pompea conclude that these ancient nautiloids saw a gigantic moon slightly more than two-fifths its current distance from the earth (yes, they had eyes).

At this point, I must confess to some ambivalence about this large body of data on fossil growth rhythms. The methods are beset with unsolved problems. How do you know what periodicity the lines reflect? Consider the case of fine lines, for example. They are usually counted as though they record solar days. But suppose they are a response to tidal cycles—a periodicity that involves both the earth's rotation and the moon's revolution. If the moon revolved in a much shorter time in the past, then ancient tidal cycles were not nearly so close to the solar day as they are now. (You should now grasp the importance of Kahn and Pompea's argument, made without direct evidence by the way, that the fine lines of *Nautilus* reflect day–night cycles of vertical migration rather than tidal effects. In fact, they explain their three exceptional cases by arguing that these nautiloids inhabited persistently shallow, nearshore waters and may have recorded the tides.)

Even if lines are a response to solar cycles, how do you assess the days per ancient month or year? Simple counting is not the solution because animals often skip a day but do not, so far as we know, double up. Actual counts generally underestimate the number of days (remember Wells's original modern corals with an average of 360, not 365, daily bands—on very cloudy days, growth during the daytime may not exceed growth at night, and bands may not form).

Moreover, to pose the most basic question of all, how can we be certain that lines reflect an astronomical periodicity at all? Too often, little beyond their geometric regularity has inspired the assumption that they record days, months, or years. But animals are not passive machines, dutifully recording astronomical cycles in all their regularities of growth. Animals have internal clocks as well, and these are often keyed to metabolic rhythms with no apparent relationship to days, tides, and seasons. For example, most animals slow down their growth rates greatly as they advance in age. But many growth lines continue to increase in size at a constant rate. The distance between septa of *Nautilus* increases constantly and regularly throughout growth. Are septa really deposited once each month, or do later ones measure longer amounts of time? *Nautilus* may live by the rule: grow a septum after reaching a regularly increasing chamber volume, not grow a septum each full moon. I am, primarily for this reason, highly skeptical about Kahn and Pompea's conclusions.

The result of these unsolved problems is a body of poorly synchronized data. Uncomfortably large differences exist in the literature. One study of supposedly lunar periodicities in corals suggests that, about 350 million years ago, the month contained three times the number of days that Kahn and Pompea would allow.

Nonetheless, I remain satisfied and optimistic for two reasons. First, despite all internal asynchrony, every study has revealed the same basic pattern—decrease in the number of days per year. Second, after an initial period of uncritical enthusiasm, paleontologists are now doing the required hard work to learn just what the lines represent—experimental studies on modern animals in controlled conditions. Criteria for the resolution of discrepancies in fossil data should soon be available.

Scarcely any geological subject could be more fascinating or more beset with juicy problems. Consider the following: if you extrapolate back through time the current recession of the moon as estimated from eclipse data, the moon enters the Roche limit about one billion years ago. Inside the Roche limit, no major body can form. If a large body enters it from outside, results are unclear but certainly impressive. Vast tides would roar across the earth and the lunar surface would melt, which, conclusively from dates on Apollo rocks, it did not. (And the recession rate estimated from modern data—5.8 centimeters per year—is much less than the average advocated by Kahn and Pompea—94.5 centimeters per year.) Clearly, the moon was not this close to us either a billion years ago or ever at all since its surface solidified more than four billion years ago. Either rates of recession have varied drastically, and were much slower early in the earth's history, or the moon entered its current orbit a long time after the earth's formation. In any case, the moon was once much closer to us, and this different relationship should have had an important effect on the history of both bodies.

As for the earth, we have tentative indications in some of our earliest sedimentary rocks of tidal amplitudes that would put the Bay of Fundy to shame. For the moon, Kahn and Pompea make the interesting suggestion that its closer position and the earth's stronger gravitational pull at that time may explain why the lunar maria are concentrated on its visible, earthward side (the maria represent vast extrusions of liquid magma), and why the moon's center of mass is displaced in an earthward direction.

Geology has no more important lesson to teach than the vastness of time. We have no trouble getting our conclusions across intellectually—4.5 billion years rolls easily off the tongue as an age for the earth. But intellectual knowledge and gut appreciation are very different things. As a sheer number, 4.5 billion is incomprehensible, and we resort to metaphor and image to emphasize just how long the earth has existed and just how insignificant the length of human evolution has been—not to mention the cosmic millimicrosecond of our personal lives.

The standard metaphor for earth history is a 24-hour clock with human civilization occupying the last few seconds. I prefer to emphasize the accumulated oomph of effects utterly insignificant on the scale of our lives. We have just completed another year and the earth has slowed down by another 1/50,000 second. So blinking what? What you have just read is what.

Bibliography

Agassiz, E.C. 1895. *Louis Agassiz: his life and correspondence.* Boston: Houghton, Mifflin.

Agassiz, L. 1850. The diversity of origin of the human races. *Christian Examiner* 49: 110–45.

Agassiz, L. 1962 (originally published in 1857). *An essay on classification.* Cambridge, Mass.: Belknap Press of Harvard University Press.

Baker, V.R., and Nummedal, D. 1978. *The channeled scabland.* Washington: National Aeronautics and Space Administration, Planetary Geology Program.

Bakker, R.T. 1975. Dinosaur renaissance. *Scientific American,* April, pp. 58–78.

Bakker, R.T., and Galton, P.M. 1974. Dinosaur monophyly and a new class of vertebrates. *Nature* 248: 168–72.

Bateson, W. 1922. Evolutionary faith and modern doubts. *Science* 55: 55–61.

Berlin, B. 1973. Folk systematics in relation to biological classification and nomenclature. *Annual Review of Ecology and Systematics* 4: 259–71.

Berlin, B.; Breedlove, D.E.; and Raven, P.H. 1966. Folk taxonomies and biological classification. *Science* 154: 273–75.

Berlin, B.; Breedlove, D.E.; and Raven, P.H. 1974. *Principles of Tzeltal plant classification: an introduction to the botanical ethnography of a Mayan speaking people of highland Chiapas.* New York: Academic Press.

Bourdier, F. 1971. Georges Cuvier. *Dictionary of Scientific Biography* 3: 521–28. New York: Charles Scribner's Sons.

Bretz, J Harlen. 1923. The channeled scabland of the Columbia Plateau. *Journal of Geology* 31: 617–49.

Bretz, J Harlen. 1927. Channeled scabland and the Spokane flood. *Journal of the Washington Academy of Science* 17: 200–211.

Bretz, J Harlen. 1969. The Lake Missoula floods and the channeled scablands. *Journal of Geology* 77: 505–543.

Broca, P. 1861. Sur le volume et la forme du cerveau suivant les individus et suivant les races. *Bullétin de la Société d'Anthropologie de Paris* 2: 139–207, 301–321, 441–46.

Broca, P. 1873. Sur les crânes de la caverne de l'Homme-Mort (Lozère). *Revue d'anthropologie* 2: 1–53.

Bulmer, R., and Tyler, M. 1968. Karam classification of frogs. *Journal of the Polynesian Society* 77: 333–85.

Carr, A., and Coleman, P.J. 1974. Sea floor spreading theory and the odyssey of the green turtle. *Nature* 249: 128–30.

Carrel, J.E., and Heathcote, R.D. 1976. Heart rate in spiders: influence of body size and foraging energetics. *Science.*

Chambers, R. 1844. *Vestiges of the natural history of creation.* New York: Wiley and Putnam.

Cuénot, C. 1965. *Teilhard de Chardin.* Baltimore: Helicon.

Darwin, C. 1859. *On the origin of species.* London: John Murray.

Darwin, C. 1862. *On the various contrivances by which British and foreign orchids are fertilized by insects.* London: John Murray.

Darwin, C. 1871. *The descent of man.* London: John Murray.

Darwin, C. 1872. *The expression of the emotions in man and animals.* London: John Murray.

Davis, D.D. 1964. The giant panda: a morphological study of evolutionary mechanisms. *Fieldiana* (Chicago Museum of Natural History) *Memoirs* (Zoology) 3: 1–339.

Dawkins, R. 1976. *The selfish gene.* New York: Oxford University Press.

Diamond, J. 1966. Zoological classification system of a primitive people. *Science* 151: 1102–04.

Down, J.L.H. 1866. Observations on an ethnic classification of idiots. *London Hospital Reports,* pp. 259–62.

Eldredge, N., and Gould, S.J. 1972. Punctuated equilibria: an alternative to phyletic gradualism. In *Models in Paleobiology,* ed. T.J.M. Schopf, pp. 82–115. San Francisco: Freeman, Cooper and Co.

Elbadry, E.A., and Tawfik, M.S.F. 1966. Life cycle of the mite *Adactylidium sp.* (Acarina: Pyemotidae), a predator of thrips eggs in the United Arab Republic. *Annals of the Entomological Society of America* 59: 458–61.

Finch, C. 1975. *The art of Walt Disney.* New York: H.N. Abrams.

Fine, P.E.M. 1979. Lamarckian ironies in contemporary biology. *The Lancet,* June 2, pp. 1181–82.

Fluehr-Lobban, C., 1979, Down's syndrome (Mongolism): the scientific history of a genetic disorder, unpublished manuscript.

Fowler, W.A. 1967. *Nuclear astrophysics.* Philadelphia: American Philosophical Society.

Fox, G.E.; Magrum, L.J.; Balch, W.E.; Wolfe, R.S.; and Woese, C.R. 1977. Classification of methanogenic bacteria by 16S ribosomal RNA characterization. *Proceedings of the National Academy of Sciences* 74: 4537–41.

Frankel, R.B.; Blakemore, R.P.; and Wolfe, R.S. 1979. Magnetite in freshwater magnetotactic bacteria. *Science* 203: 1355–56.

Frazzetta, T. 1970. From hopeful monsters to bolyerine snakes. *American Naturalist* 104: 55–72.

Galilei, Galileo. 1638. *Dialogues concerning two new sciences.* Translated by H. Crew and A. DeSalvio. 1914, New York: MacMillan.

Goldschmidt, R. 1940. *The material basis of evolution.* New Haven, Conn.: Yale University Press.

Gould, S.J. 1977. *Ontogeny and phylogeny.* Cambridge, Mass.: Belknap Press of Harvard University Press.

Gould, S.J., and Eldredge, N. 1977. Punctuated equilibria: the tempo and mode of evolution reconsidered. *Paleobiology* 3: 115–51.

Gould, J.L.; Kirschvink, J.L.; and Defeyes, K.S. 1978. Bees have magnetic remanence. *Science* 201: 1026–28.

Gruber, H.E., and Barrett, P.H. 1974. *Darwin on man.* New York: Dutton.

Günther, B., and Guerra, E. 1955. Biological similarities. *Acta Physiologica Latinoamerica* 5: 169–86.

Haldane, J.B.S. 1956. Can a species concept be justified? In *The species concept in paleontology,* ed. P.C. Sylvester-Bradley, pp. 95–96. London: Systematics Association, Publication no. 2.

Hamilton, W.D. 1967. Extraordinary sex ratios. *Science* 156: 477–88.

Hanson, E.D. 1963. Homologies and the ciliate origin of the Eumetazoa. In *The lower Metazoa,* ed. E.C. Dougherty et al. pp. 7–22. Berkeley: University of California Press.

Hanson, E.D. 1977. *The origin and early evolution of animals.* Middletown, Connecticut: Wesleyan University Press.

Hopson, J.A. 1977. Relative brain size and behavior in archosaurian reptiles. *Annual Review of Ecology and Systematics* 8: 429–48.

Hull, D.L. 1976. Are species really individuals? *Systematic Zoology* 25: 174–91.

Jackson, J.B.C. and G. Hartman. 1971. Recent brachiopod-coralline sponge communities and their paleoecological significance. *Science* 173: 623–25.

Jacob, F. 1977. Evolution and tinkering. *Science* 196: 1161–66.

Jastrow, R., and Thompson, M.H. 1972. *Astronomy: fundamentals and frontiers.* New York: John Wiley.

Jerison, H.J. 1973. *Evolution of the brain and intelligence.* New York: Academic Press.

Johanson, D.C., and White, T.D. 1979. A systematic assessment of early African hominids. *Science* 203: 321–30.

Kahn, P.G.K., and Pompea, S.M. 1978. Nautiloid growth rhythms and dynamical evolution of the earth-moon system. *Nature* 275: 606–611.

Keith, A. 1948. *A new theory of human evolution.* London: Watts and Co.

Kirkpatrick, R. 1913. *The nummulosphere. An account of the organic origin of so-called igneous rocks and of abyssal red clays.* London: Lamley and Co.

Kirsch, J.A.W. 1977. The six-percent solution: second

thoughts on the adaptedness of the Marsupialia. *American Scientist* 65: 276–88.

Knoll, A.H., and Barghoorn, E.S. 1977. Archean microfossils showing cell division from the Swaziland System of South Africa. *Science* 198: 396–98.

Koestler, A. 1971. *The case of the midwife toad.* New York: Random House.

Koestler, A. 1978. *Janus.* New York: Random House.

Leakey, L.S.B. 1974. *By the evidence.* New York: Harcourt Brace Jovanovich.

Leakey, M.D., and Hay, R.L. 1979. Pliocene footprints in the Laetolil Beds at Laetoli, northern Tanzania. *Nature* 278: 317–23.

Long, C.A. 1976. Evolution of mammalian cheek pouches and a possibly discontinuous origin of a higher taxon (Geomyoidea). *American Naturalist* 110: 1093–97.

Lorenz, K. 1971 (originally published in 1950). Part and parcel in animal and human societies. In *Studies in animal and human behavior,* vol. 2, pp. 115–95. Cambridge, Mass.: Harvard University Press.

Lurie, E. 1960. *Louis Agassiz: a life in science.* Chicago: University of Chicago Press.

Lyell, C. 1830–1833. *The principles of geology.* 3 vols., London: John Murray.

Ma, T.Y.H., 1958. The relation of growth rate of reef corals to surface temperature of sea water as a basis for study of causes of diastrophisms instigating evolution of life. *Research on the Past Climate and Continental Drift* 14: 1–60.

Majnep, I., and Bulmer, R. 1977. *Birds of my Kalam country.* London: Oxford University Press.

Mayr, E. 1963. *Animal species and evolution.* Cambridge, Mass.: Belknap Press of Harvard University Press.

Merton, R.K. 1965. *On the shoulders of giants.* New York: Harcourt, Brace and World.

Montessori, M. 1913. *Pedagogical anthropology.* New York: F.A. Stokes.

Morgan, E. 1972. *The descent of woman.* New York: Stein and Day.

O'Brian, C.F. 1971. On *Eozoön Canadense. Isis* 62: 381–83.

Osborn, H.F. 1927. *Man rises to Parnassus.* Princeton, New Jersey: Princeton University Press

Ostrom, J. 1979. Bird flight: how did it begin? *American Scientist* 67: 46–56.

Payne, R. 1971. Songs of humpback whales. *Science* 173: 587–97.

Pietsch, T.W., and Grobecker, D.B. 1978. The compleat angler: aggressive mimicry in an antennariid anglerfish. *Science.* 201: 369–370.

Raymond, P. 1941. Invertebrate paleontology. In *Geology, 1888–1938. Fiftieth anniversary volume,* pp. 71–103. Washington, D.C.: Geological Society of America.

Rehbock, P.F. 1975. Huxley, Haeckel, and the oceanographers: the case of *Bathybius haeckelii. Isis* 66: 504–533.

Rupke, N.A. 1976. *Bathybius Haeckelii* and the psychology of scientific discovery. *Studies in the History and Philosophy of Science* 7: 53–62.

Russo, F., s.j. 1974. Supercherie de Piltdown: Teilhard de Chardin et Dawson. *La Recherche* 5: 293.

Schreider, E. 1966. Brain weight correlations calculated from the original result of Paul Broca. *American Journal of Physical Anthropology* 25: 153–58.

Schweber, S.S. 1977. The origin of the *Origin* revisited. *Journal of the History of Biology* 10: 229–316.

Stahl, W.R. 1962. Similarity and dimensional methods in biology. *Science* 137: 205–212.

Teilhard de Chardin, P. 1959. *The phenomenon of man.* New York: Harper and Brothers.

Temple, S.A. 1977. Plant-animal mutualism: coevolution with dodo leads to near extinction of plant. *Science* 197: 885–86.

Thompson, D.W. 1942. *On growth and form.* New York: Macmillan.

Verrill, A.E. 1907. The Bermuda Islands, part 4. *Transactions of the Connecticut Academy of Arts and Sciences* 12: 1–160.

Walcott, C.; Gould, J.L.; and Kirschvink, J.L. 1979. Pigeons have magnets. *Science* 205: 1027–29.

Wallace, A.R. 1890. *Darwinism.* London: MacMillan.

Wallace, A.R. 1895. *Natural selection and tropical nature.* London: MacMillan.

Waterston, D. 1913. The Piltdown mandible. *Nature* 92: 319.

Wells, J.W. 1963. Coral growth and geochronometry. *Nature* 197: 948–950.

Weiner, J.S. 1955. *The Piltdown forgery.* London: Oxford University Press.

White, M.J.D. 1978. *Modes of speciation.* San Francisco: W.H. Freeman.

Wilson, E.B. 1896. *The cell in development and inheritance.* New York: MacMillan.

Wilson, E.O. 1975. *Sociobiology.* Cambridge, Mass.: The Belknap Press of Harvard University Press.

Wynne-Edwards, V.C. 1962. *Animal dispersion in relation to social behavior.* London: Oliver and Boyd.

Zirkle, C. 1946. The early history of the idea of the inheritance of acquired characters and pangenesis. *Transactions of the American Philosophical Society* 35: 91–151.

Index